GONZO GIZMOS

PROJECTS & DEVICES

TO CHANNEL YOUR INNER GEEK

SIMON QUELLEN FIELD

CHICAGO
REVIEW
PRESS

Library of Congress Cataloging-in-Publication Data
Field, Simon Quellen
Gonzo gizmos : projects & devices to channel your
inner geek / Simon Quellen Field.
    p. cm.
ISBN 1-55652-520-6
1. Electronic apparatus and appliances—Design and
construction—Amateurs' manuals. 2. Science—
Experiments. I. Title.
TK9965.F47 2004
621.381—dc22   2003014665

Cover and interior design: Laura Lindgren

Published by Chicago Review Press, Incorporated
814 North Franklin Street
Chicago, Illinois 60610
ISBN 1-55652-520-6
Printed in the United States of America
5 4 3

Contents

② ELECTROMAGNETISM

3 ELECTROCHEMISTRY

4 RADIOS

Introduction

I like to build toys.

I especially like to build toys that let me teach science.

But most of all, I like to build science toys from scraps of junk I find around the house, or from things I can buy at an office supply store or local hardware store. Things that everyone can recognize for their normal purpose but that become magic when placed in a different context.

I especially like toys that can be built in a few minutes, while the kids watch and participate, where the results come soon enough that there is less chance they will lose focus or get distracted.

When a kid first fires a Film Can Cannon, and the flash and bang send the little plastic can across the room, the squeals of delight are far out of proportion to the few minutes it took to disassemble a dead electronic lighter for the sparking mechanism, and attach a couple wires. When a child hears a radio station on a crystal radio that was a few wires and parts moments ago, the magic is evident in the wide eyes and excitement.

Science doesn't have to be difficult. I carry a few magnets with me wherever I go. Super-strong magnets are magical in their own right, but with a little bit of wire, they can become electric motors in a few minutes. What was lifeless junk a moment ago is now spinning all by itself, making little sparks and buzzing like a bee.

Not all of the projects in this book are simple. But getting 12,000 volts of electricity from a D battery makes the Van de

Graaff Generator worth the trouble. Or just being able to say you built a Plastic Hydrogen Bomb can justify the time you spend.

Most of the projects in this book go together in less than an hour, sometimes much less, and in the end you have a Gauss Rifle, or a magnet floating in midair, not touching anything, or a steam-powered boat putting around in the bathtub.

But this book is not just about making toys. It is about teaching science. In the seven years that this material has been on my Web site, I have received thousands of e-mails from kids who used the ideas in their science fair projects, or from their proud parents. These kids can stand up and explain how their project works, mostly because it was something they could build themselves, and because over seven years my answers to all their questions have been incorporated into the on-line material.

My audience is mostly adults, despite the large number of kids with science projects that are due tomorrow. The Web site is getting over 3,000 visitors per day as of this writing, and doubling every eight months. Many visitors are science teachers at all grade levels. Others are parents homeschooling their kids. Still more are tinkerers and kids at heart like me, who remember finding projects like these long ago, but never all in one place.

To all of you I say go ahead, send your voice across the street on a laser beam, build a radio from a razor blade, cook your hot dogs with solar power, and delight your friends and neighbors with amazing things even they can do, once you show them how.

Thoughts on Safety

Most of safety is about using your head.

Know which end of a soldering iron to hold onto.

If you can't trust someone with a box of matches, don't give them a Film Can Cannon.

Don't set off explosions next to your ear.

Never aim things at the cat.

Wait until things have cooled before you touch them.

Part of the fun of the following projects is in showing everyone how smart you are. Don't ruin the moment by doing something that demonstrates the opposite. Even if you never have to use it, having a fire extinguisher handy can make a project that much more exciting and impressive. Safety glasses add intrigue and a sense of danger, and everyone who wears a pair is suddenly in the cool group at the demonstration, and can stand closer.

I have left my dangerous projects out of the book. The flaming marshmallow gun was a big hit until it was banned by the treaty on unconventional weapons, and we aren't allowed to talk about trebuchets after launching Cousin Willie at the county fair. But some of the projects that did make it into the book still use matches, fuels, and magnifying glasses, or objects that can hurt if thrown in the wrong direction.

Above all, use common sense.

MAGNETISM

We've all played with magnets. A pair of magnets by itself makes a wonderful toy. Today's magnets are even better than the best ones I remember playing with as a child. At electronics and toy stores you can buy flexible plastic magnetic strips that can be cut into shapes with scissors. You can also buy inexpensive (and brittle) ceramic magnets, stronger AlNiCo magnets, and even the new super-strong rare-earth magnets. These are made of neodymium-iron-boron or samarium-cobalt, and are very powerful.

Through mail-order surplus houses, you can buy large neodymium-iron-boron magnets that are incredibly strong. These magnets, which cost about five dollars each, can hold paperback books onto your refrigerator, or drag one other around a two-inch-thick table, one on top of the table and one underneath. I once entertained my guests and several waiters at a restaurant by mysteriously moving the stainless flatware around the table. Most people are not familiar with the properties of powerful magnets and are amazed at what they can do.

Because of their high strength-to-weight ratios, neodymium-iron-boron magnets appear to be little affected by gravity. Small ones can be placed on either side of your nose, and will stay there until you laugh so hard they slide upward, against gravity, and snap together. They can also be used as temporary earrings, but be sure to handle the larger magnets with care, since they will pinch hard enough to cause blisters if they are separated by only a small bit of skin. They can also easily erase magnetically stored information on credit cards, computer floppy disks, and cassette tapes, so take care where you place them.

A Few Simple Magnet Experiments

If you are showing magnets to a very young person for the first time, choose large, cheap, ceramic magnets that will not be easily lost or swallowed. Then try a few tricks. Show how they attract and repel each other. Have the child slide donut-shaped magnets over a pencil, stacking several so that each repels the one below it, to form a magnetic spring.

SHOPPING LIST
- Cheap magnets
- Sand
- Sticky tape
- Sheet of paper

TOOLS
- Clear plastic bag
- Paper clips, pins, key chains, and other bits of ferrous metal
- Pin or screwdriver

Next, put several cheap magnets into a clear plastic bag and drag it through sand at a playground or beach. The magnets will attract black iron ore from the sand. (The plastic bags keep the ore off the magnets, keeping them clean and making it easier to remove the powder. If the powder does get onto the magnets, use some sticky tape to remove it.) Sprinkle this ore onto a sheet of paper with a magnet placed underneath, which will create arcing lines of powder that trace out the magnetic lines of force. A magnetic compass placed near the magnet will align itself with the arcs of powder, following the curve as you move it around.

Place several magnets under the paper to create interesting shapes in the iron ore above. Paper clips, pins, key chains, and other bits of ferrous metal placed beneath the paper will alter the shape of the ore powder as they become temporary magnets under the influence of the permanent magnets.

Stroke a bit of iron or steel (such as a pin or a screwdriver) across the magnet. This makes a new magnet that will attract other bits of steel or the iron ore. By placing this new magnet under the paper with the ore powder, you can demonstrate that this magnet is not as powerful as the original. Heat up the pin or screwdriver to show how heat destroys the magnetism.

Place two magnets under the paper, separated by a paper clip or other small piece of iron. Make sure each magnet touches the paper clip. Now when you sprinkle the ore on the paper, it forms a much smaller arc of powder. This shows the effect of a magnetic flux concentrator (the paper clip), which narrows and thus strengthens the force between the two magnets.

These simple experiments and observations form the backdrop for the rest of this chapter. In it you will see effects that are not so simple or obvious, but yet are simple extensions of the same properties of magnets shown above.

Magnetorheological Fluids

A magnetorheological fluid is a liquid that hardens near a magnet but becomes liquid again when you remove the magnet. These fluids are simple to make in your kitchen after a trip to a sandbox.

SHOPPING LIST
- Iron filings or beach sand
- Paper plate
- Strong magnet
- Vegetable oil

TOOLS
- Plastic bag
- Plastic spoon or other nonferrous object, such as a popsicle stick

As described earlier, mine iron ore from playground or beach sand using magnets and plastic bags. You will probably spend a while at the beach because you will need a large handful of ore. To save time, you can also purchase a vial of iron filings from a scientific supply store.

The ore you collect will likely have quite a bit of sand entrained in it. You can remove the sand with some additional refining. First, be sure the ore is dry. Spread it out on a paper plate, and hold the bag with the magnet over the plate until a small amount of ore jumps up to the bag. Put this ore onto another plate, and continue this process until no more ore rises to the magnet. Don't let the bag get too close to the plate, since there are many sand grains with ore stuck to them. You want to keep only the ore that is not stuck to grains of sand. The ore on the second plate should be visibly darker than what is left on the first plate.

If you can see a lot of sand on the second plate, repeat the process using a third plate.

Put the refined ore into a small plastic cup. The cup should be small enough so that the ore fills at least a third of it. Add a little vegetable oil to the ore, stirring the mixture with a plastic spoon or another nonferrous object, such as a popsicle stick. Keep adding oil until you get a thin black paste. Now gently place a strong magnet on the side of the cup. It should stick to the side as it attracts the ore. The ore should become quite stiff. Tip the cup over another cup to let excess oil and ore pour off. What remains in the first cup is a magnetorheological fluid.

Now you're ready for the fun part. Hold the cup upright and remove the magnet. Stir the liquid with the plastic spoon. It may be a little stiff at first, but it will soon stir easily. Tip the cup a bit to the side and bury the bowl of the spoon in the liquid. Now place the magnet on the side of the cup to stiffen the goop. The spoon will now stand upright when the cup is righted. The cup can even be inverted without losing any fluid, although a little oil may still drip out the first few times. Set the cup upright again, remove the magnet; the solid mass slumps back into the cup and the spoon falls over.

Put some of the magnetorheological fluid into a plastic bag, and stick a magnet to the outside. Now you can mold the fluid into shapes by pressing the bag. The fluid will act like clay and hold its shape. When you eventually remove the magnet, the shape will slump into a puddle.

WHY DOES IT DO THAT?

Iron ore in oil reacts pretty much the way it reacts without the oil. Only one thing is different: the oil allows the powder to slump more easily than it can when dry. This is because of the oil's additional weight, its lubricating ability, its viscosity, and the fact that the iron ore is more buoyant in oil than in air.

Saying that iron ore in oil behaves like dry ore doesn't really answer the question unless you know why dry ore acts the way it does. If you look very closely at the ore with a magnifying glass or a microscope, you will notice that the pieces are slightly longer than they are wide, like small footballs. Shapes like this do interesting things in a magnetic field.

To see why, put a small iron nail in one hand, and a magnet in the other. Move the nail around the magnet, holding the nail loosely so it can move under the influence of the magnet. The nail will align itself parallel to a bar magnet if you hold them side-by-side. But as you move the nail toward one pole of the magnet, it will rotate so that the point of the nail points toward the pole. Eventually, when the nail is above the pole, it will point directly at the pole.

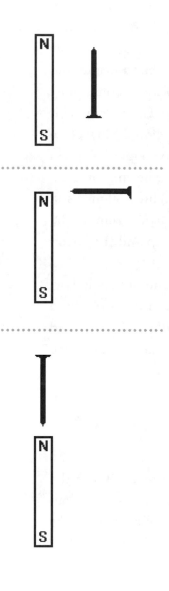

There are two ways to think about what is happening. Pretend the attraction of the magnet is like the gravitational attraction of the Earth and that the nail is a domino standing up, on its end. A slight push makes the domino lie flat, and it takes a larger push to make it stand straight up again. Physicists say that the domino has more potential energy when it is standing up than it has when it is lying down. There is a tendency for a domino to lose this energy by lying down. It has much less of a tendency to spontaneously gain energy and stand back up. Similarly, the nail tends to "fall down" so that it aligns with the magnetic lines of force that surround the magnet.

The second way of thinking about the nail and magnet requires a second nail. Hold the first nail parallel to the magnet, about half an inch to the right. Bring the second nail parallel to the first nail, a little to its right. You might expect the magnet to attract the second nail, just like the first nail did, but instead you'll find that the two nails *repel* one another. If you lower the second nail so its top is near the bottom of the first nail, it now *attracts* the first nail.

The nails seem to have become magnets themselves while in the presence of the bar magnet. Their poles repel when they are parallel, and attract when they align vertically. If the bar magnet has its north pole facing away from you, the nails will have their south poles facing away from you. The nails attract the magnet because unlike poles attract each other, but the nails repel each other because like poles repel. It is now easy to see why the nail follows the magnetic lines of force as it moves around the magnet: its north pole points toward the magnet's south pole, and its south pole points to the magnet's north pole. When the nail is beside the magnet, it rests parallel because these two attractions are equal. When the nail is closer to the magnet's north pole, the nail's south pole attracts and the north pole repels, so the nail rotates.

Magnets repel one another when their matching poles are side-by-side. They attract one another when their opposite poles are end-to-end. The natural state of a collection of magnets will thus be a string of them stuck end-to-end. If there are two strings next to each other, they will stagger so the poles of one string will be next to the center of the other string. In this arrangement, like poles are as far apart as possible. These strings will still repel from one another slightly, which is exactly the behavior of grains of iron ore sprinkled on paper above a magnet.

When a magnet is placed on the side of a jar of iron ore powder, the powder arranges itself into strings. Each grain of powder becomes a magnet and attracts the opposite pole of its neighbor. The strings thus formed repel each other, and the powder expands. If the powder has been mixed with oil, the oil wicks into the spaces created by the expansion and sticks there by surface tension. The result is a dry-appearing solid that does not leak oil.

A Magnetic Heat Engine

SHOPPING LIST
- Canadian nickel or rare-earth magnet
- Stiff copper or brass wire
- Large ceramic magnet
- Small candle or alcohol burner

TOOLS
- Wire coat hanger (optional)
- Needle-nosed pliers

A simple Magnetic Heat Engine demonstrates how heat causes magnetic material to lose its ability to be magnetized, and how cooling that material allows it to regain its magnetic property.

I originally built this device using a Canadian nickel. Canadian nickels are made of pure nickel, unlike U.S. nickels, which contain so much copper that they are not magnetic. You can build this device with the nickel or with a rare-earth magnet. A rare-earth magnet will work a little better because it loses its magnetic properties at a lower temperature, and thus the engine can use a candle instead of an alcohol burner for its heat source.

The pendulum on the Magnetic Heat Engine should only swing back and forth in one direction, so it needs two wires to suspend it. Cut about a foot of stiff copper or brass wire and wrap the center of the wire around the large ceramic magnet. Then, using needle-nosed pliers, twist both ends into small loops. Bend the loops up to form the two supports for the pendulum on page 10.

If you are using the rare-earth magnet for the pendulum's weight, it helps at this point to demagnetize it by holding it in a candle flame. You can stick it onto a coat hanger and hold the magnet in the flame until it falls off. Demagnetizing the rare-earth magnet will prevent it from jumping onto the large ceramic magnet while you adjust the pendulum.

Wrap another foot of wire around the nickel or the rare-earth magnet that will act as the pendulum weight. Bend the two ends of the wire to hang on the loops of the pendulum support, making sure that the pendulum weight is just close enough to the magnet that it rises to it when the pendulum is vertical. The wires of the pendulum and its support should be long enough that the weight can fall away from the flame and hang vertically when it is demagnetized. Cut off the remaining wire.

Suspend the Canadian nickel (or rare-earth magnet) at the end of a pendulum. Place a large magnet near the pendulum, so that the nickel sticks out toward the large magnet. The magnet should be close enough that the nickel hangs at an angle toward the magnet, and not straight down, at the bottom point of the pendulum's swing.

If you have chosen to use the Canadian nickel, you will need a better heat source than a candle. A small alcohol lamp or fondue pot burner will do nicely. You may have to make the pendulum support wires longer to make room for the lamp.

Place the alcohol lamp or candle under the nickel, so the flame just touches it. The flame will heat up the nickel until it loses its ability to be magnetized. Gravity will eventually pull it away from the large magnet (and thus away from the flame). The nickel will then cool down once it is away from the flame, regaining its ability to be magnetized. The large magnet will then pull it up into the flame, and the whole process will repeat.

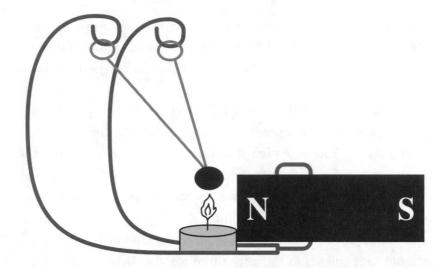

If the weight still touches the flame when it has fallen away from the magnet, adjust the pendulum's supports so that the weight rests a little farther away. Be careful when adjusting the supports, since they may be quite hot. (Also, be careful to move the heat source so as not to burn yourself on the flame.) If the weight is so far away that the magnet cannot pull it back up once it is magnetized, adjust the supports to bring it closer. When the engine is adjusted just right, it will settle down to a predictable swing, often taking only one swing to cool enough to stick to the magnet again. It will run as long as the flame burns.

WHY DOES IT DO THAT?

A heat engine works because of something called the Curie effect. The Curie effect describes how a magnetic material loses its ability to stick to a magnet when heated above a certain temperature. This temperature is called the Curie temperature, and varies with the material.

The Curie temperature for iron is about 800° Celsius (C). The Curie temperature for the inexpensive ceramic magnets is also quite high, which is why the candle flame or even the alcohol lamp does not affect them. The Curie temperature for the Canadian nickel is lower, about 631°C. This temperature is within range of an alcohol lamp, and almost possible with a candle. The Curie temperature for a rare-earth magnet is 310°C, and the candle can reach this easily, not only because of the lower Curie temperature but because the magnets are so much smaller than the nickel that they heat up faster and have less unheated surface area.

I have tried Ronson lighter flints, which also have a Curie temperature within easy range of a candle flame. The combination of their small size and low Curie temperature makes them stay above their Curie point too long. The magnet and the flame have to be close together for the engine to work. When the flints are close enough to the magnet to overcome gravity, they will be close enough to the flame to rise above their Curie point.

Other heat engine designs are possible, involving placing the flints on a wheel and using a soldering iron as a heat source. A magnifying glass could also be used to focus the radiant energy of the sun on a flint when it is touching the magnet, but not heating the flint when it falls away. Experiment with other designs; there are many possibilities.

MORE ABOUT MAGNETS
(The Scientific Part)

It is fairly simple to visualize nails becoming magnets and aligning themselves in the magnetic field. It is also easy to visualize them as dominoes that "fall" in line with the field. But what is really happening?

Magnetic fields exist because somewhere electrons are moving. In the nails, the electrons in the iron atoms are orbiting the nuclei, and each electron is also spinning on its axis. Most materials are not strongly magnetic because their magnetic poles (caused by spinning electrons) are either randomly oriented or they pair in opposite directions.

If the electrons in an atom (or a molecule) create equal numbers of north and south magnetic poles, they cancel each other out and the atom has no magnetic poles. If a magnet is brought near such an atom, the magnetic field causes the electrons in the atom to move. The moving electrons create a magnetic field, and the field they create is opposite to the original magnetic field. The atoms move away from the magnet. Materials that act this way are called *diamagnetic* materials. The effect is very weak, even in bismuth, which is the most diamagnetic material available.

If the electrons in an atom do not create canceling magnets, the atom has a north and a south pole. Normal room temperature causes these atoms to bounce around, and their magnetic poles are randomly oriented. If the atoms are inside a magnetic field, some of them will align with the field. Others will have too much energy and will continue to bounce around. The amount of magnetic energy created depends on the atoms' temperature; the colder the atoms are, the more of them will align with the field. Materials that act in this way are called *paramagnetic* materials. Paramagnetic materials are attracted to magnets but the effect is weak unless the temperature is very low. Paramagnetic materials also exhibit diamagnetic effects, but the paramagnetic effect is stronger

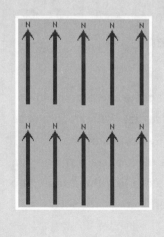

unless the temperature is very high. Aluminum and platinum are paramagnetic.

In four elements—iron, nickel, cobalt, and gadolinium—the magnetic poles of the atoms stay lined up even at temperatures that thoroughly randomize the poles of other elements. These elements are called *ferromagnetic.*

In ferromagnetic elements, the magnetic effect is caused by the electrons in the inner orbits. They would align with their poles opposite except for a principle of quantum mechanics called the Pauli Exclusion Principle. This principle says that electrons cannot occupy the same space unless they have opposite spins, which means they have opposite magnetic poles. For example, two atoms in an element like iron might share an outer electron. This electron will try to align the poles of the inner electrons to be opposite. The inner electrons end up having the same orientation. Thus aligned, they contribute to the magnetic field.

In many paramagnetic materials, the atoms pair up with their north and south poles in opposite directions, canceling each other. These materials are called *antiferromagnetic*. Chromium and manganese are antiferromagnetic.

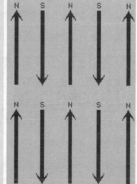

Some materials, called *ferrites*, are similar to antiferromagnetic materials but consist of two different magnetic components, one of which is stronger than the other. This makes them act like ferromagnetic materials, and they are attracted to magnets. The iron ore used in the previous experiments is an example of a ferrite. The mineral is called magnetite; a natural magnet, lodestone, is made out of it. The name for this configuration is ferrimagnetism, after the ferrites. Ferrites find many uses in electronics because they are magnetic but do not conduct electricity like iron does.

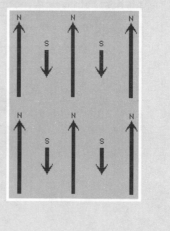

Just as electricity travels through some materials better than others, magnetism travels with ease through some materials but has more difficulty traveling through others. The ability of a material to conduct magnetic flux is called the *relative permeability* of the material. It is measured relative to the permeability of air, which is assigned a baseline permeability of 1. The relative permeability of a material can be thought of as the ease with which a material can be magnetized. The higher the relative permeability, the easier it is to magnetize a material. Iron and steel have relative permeabilities between 100 and 9,000. Ferrites have relative permeabilities between 50 and 3,000.

The lines of force of a magnet, as a group, make up the *magnetic flux*. The stronger the magnet, the more lines of force it has, and thus the more magnetic flux it has. A typical household magnet has between 100 and 1,000 lines of force. One line of force is called a *maxwell*. Materials with high relative permeability allow more lines of flux inside them than does the air around them. The flux is therefore concentrated in the material. The measure of concentration of flux, measured in lines of force per square centimeter, is the *gauss*. One gauss is one line per square centimeter. The earth's magnetic field measures about 0.2 gauss. The superconducting electromagnets used in magnetic resonance imaging (MRI) range from 5,000 to 15,000 gauss.

When you hold a nail close to a magnet, the lines of force of the magnet flow more easily through the nail than through the air. Just as electrcity seeks the path of least resistance, magnetic flux seeks out the highest relative permeability. Because there is less resistance if the flux flows through an entire nail lengthwise, rather than through a small part of it crosswise, there is a force on the nail that tends to line it up with the lines of force of the magnet. This is the force that makes it "fall down" parallel to the lines of force. At the same time the nail is magnetized, and its poles are attracted to the opposite poles of the magnet.

◎　◎　◎

MEASURES USED IN PERMANENT MAGNETS

What makes a good permanent magnet? A magnet should have a strongly concentrated magnetic field. It should also stay magnetized. To understand how we measure a permanent magnet's strength and permanence, it is helpful to picture how a magnet is made.

The process starts with a piece of iron or another ferromagnetic material. (It is assumed to not be previously magnetized.) To magnetize the material, it is placed in a magnetic field created by another magnet. This is often done in a field created by an electromagnet so that the strength of the field can be varied to see what happens to the iron.

When the field strength is small, the number of lines of force concentrated in the iron is also small. As the field strength rises, more of the tiny magnets in the iron, called *domains*, align with the external field. The number of lines of force in the iron rises with the field strength. This continues until all the magnets in the iron line up with the external field. At this point the iron is *saturated*—it cannot hold any more lines of force. As the external field strength rises, there is no further increase in the number of lines of force in the iron.

Suppose you now start reducing the external field strength. The heat inside the iron at room temperature jostles the tiny magnetic domains inside, and some of them readjust so they do not align with the external field. Still, some of the domains remain "stuck," aligned with the field. When the external field is completely removed, these stuck domains are left, and you have a permanent magnet.

Suppose you now reverse the external magnetic field and start gradually increasing its strength. At first, the stuck domains resist turning around to align with the external field, and the number of lines of force concentrated in the iron changes only slowly. As the external field gradually overwhelms these stuck domains, they flip over. Eventually the external field is strong enough to flip them all over, and the iron is saturated again, but the poles are now reversed.

A graph of the number of lines of force in the iron against the strength of the external magnetic field will look like a fat S.

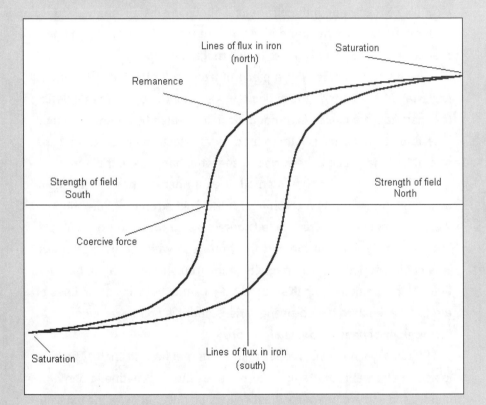

The magnetic flux density that remains in the iron when the external field is removed is the *remanence*. The strength of the external field needed to demagnetize the iron is called the *coercive force*.

The best permanent magnet has a high remanence (measured in gauss) and a high coercive force (measured in oersteds). You can buy small magnets made from a neodymium-iron-boron alloy that have very high values for remanence and coercivity. The package lists the remanence (which is often called by its other name, *residual induction*) of 10,800 gauss. The coercivity is 9,600 oersteds.

Another measurement used to describe permanent magnets is the strength of the external field when the material is saturated multiplied by the number of lines of flux in the material at saturation. This number is the *energy product*, or the *peak energy density*, measured in gauss-oersteds. Magnets from Radio Shack have a value of 27 million gauss oersteds. These magnets could be dangerous if they weren't so small. They are less than a quarter of an inch in diameter and about a tenth of an inch high.

◎   ◎   ◎

A Levitating Magnet

Floating in midair between two metal plates, a tiny magnet bobs and spins in the wind from the viewer's breath. No batteries are used, no electromagnets, no supercooled superconducting materials, just some easy-to-obtain materials from local stores.

SHOPPING LIST

- 12 ceramic ring magnets
- Tiny neodymium-iron-boron magnet
- 8 inches of ¼-inch threaded brass rod
- 4 brass hex nuts to fit the ¼-inch rod
- 2 brass or nylon washers to fit the brass rod
- Wooden base (5-inch square, ¾ inch high)
- 6–8-inch wooden support, 2 inches square
- 5 x 2-inch wooden block, ¾ inch thick
- 2 tablespoons of bismuth (nontoxic bird shot, from a gun shop or sporting goods store, or from www.scitoys.com)
- Titebond wood glue (or similar brand)
- 5-minute epoxy
- Large, cheap cooking spoon (it will be ruined)
- Empty aluminum soda can
- Wire coat hanger

TOOLS

- ¼-inch drill bit and drill
- Coarse sandpaper or large metal file

The heart of this device is the bismuth plates, which have special magnetic properties. These are what make the magnetic suspension possible.

Bismuth is very similar to lead. It is easy to melt and is as heavy and hard as lead. Because lead is toxic and bismuth is not (it is the active ingredient in Pepto-Bismol, a medicine for upset stomach), bismuth is used in environmentally safe bird shot for hunters. It is easily found in this form at gun shops and sporting goods stores. Bismuth is also used in some fishing lures as a replacement for lead.

Melting and Casting the Bismuth

The first step in assembling a levitating magnet is to melt bismuth and pour it into a mold to make the two suspension plates. The bottom of an aluminum soda can is used as a mold—it has a convenient indentation that is about the right size. You can use other molds if you wish, such as aluminum muffin tins, but realize that they will not be suitable for cooking afterward.

Place about a tablespoon of bismuth shot into the cheap spoon purchased for this purpose—don't use your best kitchen spoon! Heat the shot in the spoon over a stove or propane torch until it melts. Use a bit of coat hanger wire at this point to rake the slag (bismuth oxides) off of the top of the melted metal and onto the side of the spoon.

Next, carefully move the spoon over to the upside-down aluminum can, and pour the metal into the hollow in the bottom of the can.

Let the metal cool until it hardens, and the can is cool enough to hold. Only after the metal is completely solid can it be held under running water and pried out of the mold.

Use sandpaper or a metal file to make the top and bottom surface of the metal nice and flat.

Then, make another bismuth disk like the first one.

Building the Support Structure

Drill a ¼-inch hole in one end of the 5 x 2-inch block, about ¾ of an inch from the edge. This will accept the brass rod.

Glue the 6-inch 2 x 2 support block to the base.

Glue the 5 x 2-inch block to the top of the support block, leaving the end with the hole free. Allow the glue to dry. If you are very sparing with the glue, using only a very thin layer, the drying will only take a few minutes.

Assembling the Magnet Stack

Insert the brass rod into the hole in the wood block. Thread one nut onto the rod from above, and two from below.

Place a washer onto the rod from below, followed by the 12 magnets, and another washer, then the final nut. The last nut should be close to the end of the rod. Tighten it by hand so as not to shatter the ceramic magnets. Don't worry yet about getting the top two nuts tight around the wood block, since you will be adjusting the height of the magnet soon.

Assembling the Bismuth Plates

The levitator is very sensitive to three things: (1) the distance between the two bismuth disks, (2) the height of the magnet above them, and (3) the magnet's strength.

The disks are kept apart by three spacers. These can be made from three pieces of bismuth shot, although wood or plastic would work fine. The spacing between the disks should be very close, leaving only a very small space for the magnet. If the distance is too large, the magnet will jump from the bottom to the top, or fall from the top to the bottom, instead of floating between the disks in midair. If the magnets are very strong and far away, the distance between the bismuth disks can be increased.

Using the 5-minute epoxy, glue the three pieces of shot to the bottom bismuth disk, as far away from each other as possible, in an equilateral triangle.

Once the epoxy has completely hardened (give it 10 minutes to be safe), start filing or sanding the tops of the shot to make them just tall enough so the tiny neodymium magnet can fit with only a little room to spare. If you stack two of the tiny magnets together, they should just fit, touching both plates (only one magnet will actually be used).

Adjusting the Magnets for Levitation

Hold the top bismuth disk in one hand, just under the stack of ring magnets. Let the small neodymium magnet stick to the bottom of the bismuth disk, attracted to the ring magnets.

Now place the top bismuth disk onto the three supports on top of the bottom disk. The tiny magnet may drop to the bottom disk, or it may remain stuck to the top disk.

If the tiny magnet falls to the bottom disk, lower the ring magnet assembly slowly by turning the rod while holding the top nut. If the tiny magnet remains stuck to the top disk, raise the ring magnet assembly by turning the rod and holding the top nut.

At one point, while turning the rod, the magnet will start to levitate between the disks. The levitating disk is very sensitive to slight changes in the height of the ring magnet assembly. The rod must be turned very slowly to make tiny adjustments to the height of the neodymium magnet.

Troubleshooting

The most common problem in building the device is that the floating magnet will either stick to the top bismuth plate or stick to the bottom bismuth plate, but never hover in between them.

This happens because the plates are too far apart for the strength of the top magnet to overcome. You can either put the plates closer together or, better yet, make the top magnet stronger by adding more magnets, using more powerful magnets, or both.

The stronger the top magnets, the farther they can be from the floating magnet and still lift it. As the top magnet gets stronger and farther away from the floating magnet, the area where the floating magnet can be stabilized increases. The magnetic field is bigger and spreads out more, and so the floating magnet can move up and down more before the strength of the field changes enough to make it fly up or fall down. Scientists say the magnetic field *gradient* is weaker if the top magnet is farther away. The gradient is how much the strength of the field changes with distance.

The picture at right of a slightly different model, where a brass wire was embedded into the bismuth before it cooled. This is somewhat more difficult to adjust than the first model, since the brass wire is springy, but it works pretty well. The floating magnet is a tiny gold-plated cube of neodymium-iron-boron supermagnet, so the faces of the cube catch the light as it rotates.

🤖 WHY DOES IT DO THAT? 🤖

The bismuth disks are diamagnetic. This means that they push away from a magnet. It doesn't matter whether the north pole of the magnet or the south pole is used, the bismuth always pushes away. But diamagnetism is very weak—even in bismuth, which has the strongest diamagnetism of any metal. This is why the adjustment on the device is so sensitive.

The ring magnets attract the small neodymium magnet with just the right amount of force to counteract gravity. However, if the bismuth disks were not there, pushing it back downward, the tiny magnet would jump up to the ring magnets because as it gets closer, the force is stronger.

Just at the critical point where the magnetic pull upward just barely counteracts gravity downward, the weak diamagnetism of the bismuth will keep the magnet from jumping up to the ring magnets or falling down. The magnet floats, being repelled by the top and bottom bismuth disks.

Levitating Pyrolytic Graphite

Some materials are more diamagnetic than bismuth. These include superconductors (which at this time require cryogenic temperatures to work) and similar materials that exhibit "giant diamagnetism" (also at very low temperatures).

But there is one material that is more diamagnetic than bismuth at room temperature, at least in one direction. That material is called pyrolytic graphite. Pyrolytic graphite is a synthetic material, made by a process called chemical vapor deposition. To make pyrolytic graphite, methane gas at low pressure (about 1 Torr) is heated to 2,000°C. Very slowly, at one thousandth of an inch per hour, a layer of graphite grows.

Graphite made this way is very highly ordered, and the layers of carbon atoms form like a crystal of hexagonal sheets. These sheets lie on top of one another like sheets of mica. You can separate the layers with a sharp knife to make thinner sheets.

Pyrolytic graphite is more diamagnetic than bismuth, but only in the direction perpendicular to the sheets of carbon. In other directions it is still diamagnetic but not as strong as bismuth.

Because the density of pyrolytic graphite is lower than bismuth—its specific gravity is 2.1—it is light enough to be levitated above a sufficiently powerful magnet. A thick piece will be too heavy, since the material above about the first ½ millimeter does not contribute much to lift. But if the piece is thin enough, about ½ millimeter thick, it will simply rise up off of a strong neodymium-iron-boron supermagnet and refuse to sit still on it.

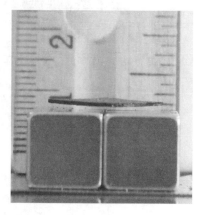

SHOPPING LIST
ͻ Pyrolytic graphite plate
ͻ Four magnets

TOOLS
☐ Sharp knife

To make the pyrolytic graphite plate hover above a magnet, it must be forced toward the center to keep it from sliding off. This can be done using four magnets. The poles of the magnets push on the diamagnetic material more strongly than other parts of the magnet. The four edges of the square of pyrolytic graphite will be pushed away from the four poles of the magnets. If the graphite square is slightly smaller than half the width of the four magnets (a little smaller than one magnet), then you can place it in the center, and it will be pushed to the middle and stay.

Since diamagnetic materials are repelled by either pole on a magnet, you can place the magnets with alternating north and south poles, and they will stick nicely to one another. If you place the whole array on a piece of sheet steel, the magnets will stay put.

Pyrolytic graphite is not easy to find. The local hardware store won't be carrying it any time soon. But you can get it at www.scitoys.com, in pieces just the right size for the levitating magnet project. These pieces are 16 millimeters wide, 32 millimeters long, and between a ½ millimeter and 1 millimeter thick. Split into thin sheets and then cut in half with a sharp knife to make 16-millimeter squares, they are perfect for levitating above four magnets.

The knife you use should be very thin so it splits the graphite cleanly without breaking it. A razor blade or a utility knife would work better than the small blade of a Swiss army knife, but an army knife works if you're very careful and adept at using it. Place the blade carefully in the middle of the edge of the graphite. Slowly push the blade in with a slight rocking motion. The graphite will make a nice clean sound as it starts to split.

Sometimes you will end up with one thin piece and one thicker piece after they are split. You can often split the thicker piece again, giving you three pieces. If you are very skilled you can get four pieces, but you will probably break a few while gaining that skill.

Once the slices are very thin, cut them in half by rocking the sharp knife over the middle of each one. The pieces will snap and may fly some distance unless you put a finger over the sides to hold them down.

 Place the thinnest sheet you have above the magnets. The thicker graphite pieces will be too heavy to float on the magnets, but a nice thin sheet will float, and a very thin sheet will float the highest. While the pyrolytic graphite sheet is floating above the magnets, push it down with your finger and watch it spring back.

Try This

Since pyrolytic graphite is a little more diamagnetic than bismuth, it makes a great substitute for bismuth in the levitating magnet project.

The picture at right shows a magnet spinning between two pieces of pyrolytic graphite, separated by a small piece of wood. The large magnet above it is not shown, but this is the same device you built using bismuth.

The Gauss Rifle:
A Magnetic Linear Accelerator

This very simple device uses a magnetic chain reaction to launch a steel marble at a target at high speed. The device takes only a few minutes to build and is simple to understand yet fascinating to watch and to use.

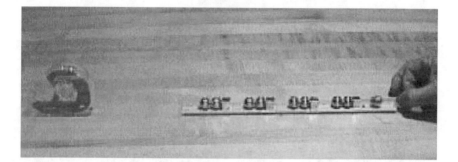

SHOPPING LIST

⊃ Nine steel balls
⊃ Sticky tape
⊃ Four magnets
⊃ Wooden ruler that has a groove in the top

TOOLS

◻ Sharp knife

The materials are simple. You need a wooden ruler that has a groove in the top in which a steel ball can roll easily. Any piece of wood or aluminum or brass with a groove will work, but rulers are easy to find.

You also need some sticky tape. Again, almost any kind will do. The photo shows transparent tape, but vinyl electrical tape works just as well.

Next, you need four magnets. Almost any type will do, but the stronger the magnets are, the faster the balls will move. Super-strong, gold-plated, neodymium-iron-boron magnets are shown (and are available at www.scitoys.com).

Last, you need nine steel balls, all with a diameter that closely matches the height of the magnets. A ⅝-inch-diameter nickel-plated steel ball works well.

The only other tool you need is a knife for trimming the tape.

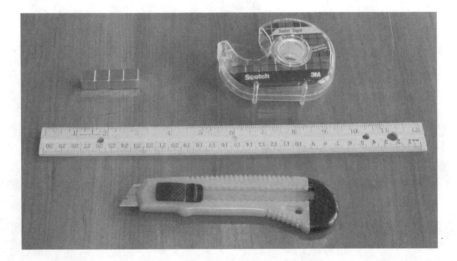

Start by taping the first magnet to the ruler at about the 2.5-inch mark. The distance is somewhat arbitrary; you simply need to get all four magnets on a 1-foot ruler. Feel free to experiment with the spacing.

With the sharp knife, trim off any excess tape. Be careful because the knife will be strongly attracted to the magnet. It is also

very important to keep the magnets from jumping together. They are made of a brittle sintered material that shatters like ceramic. Tape the ruler to the table temporarily so that it doesn't jump up to the next magnet as you tape the second magnet to the ruler.

Continue taping the magnets to the ruler, leaving an equal distance between each one.

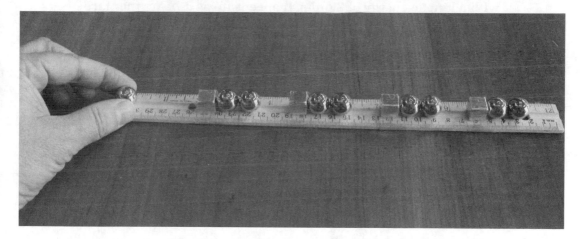

When all four magnets are taped to the ruler, it is time to load the Gauss Rifle with the balls.

To the right of each magnet, place two steel balls. Arrange a target to the right of the device so the ball does not roll down the street and get lost.

To fire the Gauss Rifle, set a steel ball in the groove to the left of the leftmost magnet. Let the ball go. If it is close enough to the magnet, it will start rolling by itself and hit the magnet.

Be ready. When the Gauss Rifle fires, it will happen too fast to see. The ball on the right will shoot away from the gun and hit the target with considerable force. This 1-foot-long version is designed so that the speed of the final ball is not enough to hurt someone, and you can use your hand or foot as a target.

🤖 WHY DOES IT DO THAT? 🤖

When you release the first ball, it is attracted to the first magnet. It hits the magnet with a respectable amount of force, with a kinetic energy we will call "1 unit." The kinetic energy of the ball is transfered to the magnet, then to the ball that is touching it on the right, and then to the ball touching that one. This transfer of kinetic energy is familiar to billiards players—when the cue ball hits another ball, the cue ball stops and the other ball speeds off.

The third ball is now moving with a kinetic energy of 1 unit. But it is moving toward the second magnet. It picks up speed as the second magnet pulls it closer. When it hits the second magnet, it is moving nearly twice as fast as the first ball did.

The third ball hits the magnet, and the fifth ball starts to move with a kinetic energy of 2 units. It speeds up as it nears the third magnet, and hits with 3 units of kinetic energy. This causes the seventh ball to speed off toward the last magnet. As it gets drawn to the last magnet, it speeds up to 4 units of kinetic energy. The kinetic energy is now transferred to the last ball, which speeds off at 4 units to hit the target.

Another way of looking at the mechanism is that at the moment the device is all set up and ready to be triggered, there are four balls that are touching their magnets. These balls are at what physicists call the "ground state." It takes energy to move them away from the magnets.

But each of these balls has another ball touching it. These second balls are not at the ground state; they are each $\frac{5}{8}$ of an inch from a magnet. They are easier to move than the balls that are touching the magnet. If you were to take a ball that was touching a magnet and pull it away from the magnet until it was $\frac{5}{8}$ of an inch away, you would be adding energy to the ball. The ball would be pulling toward the magnet with some considerable force. You could get the energy back by letting the ball go.

But after the Gauss Rifle has fired, the situation is different. Now each of the balls is touching a magnet. There is one ball on each side of each magnet. Each ball is in its ground state and has given up the energy that was stored by being ⅝ of an inch from a magnet. That combined energy of the moved ball has gone into the last ball, which uses it to destroy the target.

The kinetic energy of an object is defined as its mass times the square of its velocity. As each magnet pulls on a ball, it adds kinetic energy to the ball linearly. But the speed does not add up linearly. If you have four magnets, the kinetic energy is 4, but the speed goes up as the square root of the kinetic energy. As you add more magnets, the speed goes up by a smaller amount each time. But the distance the ball will roll, and the damage it causes to what it hits, are functions of the kinetic energy and thus functions of how many magnets we use.

You can keep scaling up the Gauss Rifle until the kinetic energy gets so high that the last magnet is shattered by the impact. After that, adding more magnets will not do much good.

Would a circular track be a perpetual motion device?

Many wonder what would happen if the track were circular. Would it create free energy? Would the balls keep accelerating forever?

I have been tempted to reply with the famous quote: "There are two kinds of people in the world—those who understand the second law of thermodynamics, and those who don't." However, I am not the kind of person to leave an inquiring mind unsatisfied, and it is more productive to explain in a little more depth what is going on.

Suppose you made a circular track and put two balls after each magnet. When the last ball is released, it encounters a magnet that has two balls at the ground state. There is no energy to be had from this magnet. The ball just bounces back.

Now suppose you had placed three balls after each magnet. When the last ball is released, it hits a ball that is ⅝ inch from the magnet. It has not gained much momentum, because most of the momentum gained is in the last half-inch as the magnet pulls more strongly on things that are closer. But the ball has enough energy from previous accelerations to release the next ball. However, that ball has less energy than the ball that caused it to release. It may have enough energy to release another ball or two, but each ball that is released has less energy than before, and eventually the chain stops.

You can show by inductive logic that no matter how many balls you stack in front of each magnet, eventually the system stops. To estimate the losses due to heating the balls as they compress when hit, consider a plastic tube standing upright on a table. Place one steel ball at the bottom of the tube. Now drop another ball into the tube so it hits the ball at the bottom and bounces back up. Measure how high the ball bounced. If it bounces halfway back up, the losses are 50 percent. Perform the experiment for yourself with the balls from the Gauss Rifle. How high does your ball bounce?

2

ELECTROMAGNETISM

Back in the 1960s my father taught me how to make the little electric motor you will make and modify in this chapter. Sometime in the 1980s, I saw a description of it in the magazine *Physics Teacher*. Lately I have seen it described as Beakman's Motor, after the science-oriented TV show on which it appeared. The device has always been the same, and it can be built in 10 minutes.

A Quickie Electric Motor

An electric motor can be built with a battery, a magnet, and a small coil of wire you make yourself. There is a secret to making it that is at the same time clever and delightfully simple.

SHOPPING LIST

- ↄ Battery holder (RS #270-402 holds a C cell; RS #270-403 holds a D cell)
- ↄ Battery to fit the holder
- ↄ Magnet (such as RS #64-1877, #64-1895, #64-1883, #64-1879, or #64-1888)
- ↄ 1 yard of enamel-coated 22-gauge (or thicker) magnet wire (RS #278-1345)
- ↄ 1 foot of bare 18- or 20-gauge wire (RS #278-1217 or #278-1216)

TOOLS

- ▢ Ballpoint pen or AAA battery

Start by winding the *armature*, the part of the motor that moves. To make the armature nice and round, wind it on a cylindrical coil form, such as a ballpoint pen or a small AAA battery. The diameter is not critical but should be related to the wire size. Thin wire requires a small form; thick wire requires a larger form.

Leaving a couple of inches of wire free at one end, wind 25 or 30 turns around the coil form. Don't try to be neat; a little randomness will help the bundle keep its shape better. The coil will end up looking something like the one at right:

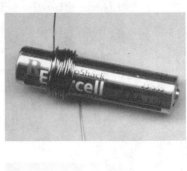

Carefully pull the coil off of the form, holding the wire so that it doesn't spring out of shape. To make the coil hold its shape permanently, wrap each free end of the wire around the coil a couple of times. Make sure that the new binding turns are exactly opposite each other so the coil can turn easily on the axis formed by the two free ends of wire, like a wheel.

It is not necessary, but I usually wrap a few turns around these binding turns as well, threading the wire into the space between the large coil and the small coils that hold it together. This makes for a neat, tight package, as in the photo at right:

If this method of holding the coil together is too difficult, feel free to use transparent tape or electrical tape to do the job. It's important that the coil be kept together, that you have the two ends of the wire anchored well and aligned in a straight line, to form a good axle.

Here is where the secret comes, the trick that makes the motor work. It is a small and subtle thing, and is very hard to see when the motor is running. Even people who know a lot about motors may be puzzled until they examine it closely. Hold the coil at the edge of a table so the coil is straight up and down (not flat on the table) and one of the free wire ends is lying flat on the table. With a sharp knife, remove the top half of the insulation from the free wire end. Be careful to leave the bottom half of the wire with the enamel insulation intact. The top half of the wire will be shiny bare copper, and the bottom half will be the color of the insulation, as shown at right:

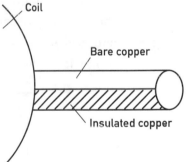

Do the same thing to the other free wire end, making sure that the shiny bare copper side is facing up on both wire ends.

The idea behind the trick is that the armature is going to rest on two supports made of bare wire. These supports will be attached to each end of the battery, so electricity can flow from one support into the armature and back through the other support to the battery. But this will only happen when the bare half of the wire is facing down, touching the supports. When the bare copper half is facing up, the insulated half is touching the supports, and no current can flow.

Next, make the axle supports. These are simple loops of wire

that hold up the armature and allow it to spin. They are made of
bare wire so that they can transmit electricity to the armature.
Take a stiff piece of bare wire—copper or brass will work, as will
a straightened paper clip—and bend it around a small nail to
make a loop in the middle, as shown below. Do the same to another
wire, so you have two supports.

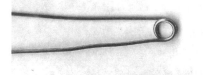

The base for this quickie motor will be the
battery holder. It makes a nice base because
it is heavy when the battery is installed (so
the motor won't wobble) and because it
has convenient holes in the plastic where you can attach the
bare wire armature supports.

Attach the support wires securely to the battery holder by
winding the free ends several times through the small holes in
the plastic at each end. Bend the support wires so the rings are
just far enough apart for the armature to spin freely. Bend them
apart a little and insert the armature into both rings, then bend
them back so they are close to the coil but not touching it.

Insert the battery into the holder. Place the magnet on top of
the battery holder just underneath the coil. Make sure the coil
can still spin freely and just misses the magnet. The finished
motor looks like this:

Note that there is a strip of paper stuck between the battery and
the electrical contact in the holder. This is the on/off switch. Remove
the paper to allow electricity to flow into the motor, and replace
the paper when you want to stop the motor and save the battery.

Spin the armature gently to get the motor started. If it doesn't start, try spinning it in the other direction. The motor will only spin in one direction.

If the motor still doesn't start, carefully check all the electrical connections. Is the battery connected so one support touches the positive end of the battery and the other touches the negative end? Is the bare copper half of the armature wire touching the bare support wires at the bottom, and only at the bottom? Is the armature freely spinning? If all these things are correct, your little motor should be spinning at a fast rate.

Try holding it upside down. The motor should spin in the opposite direction if the magnet is on top instead of on the bottom. Try turning the magnet upside down and see which direction the motor spins.

If you want a motor that has the magnet on the side instead of the top or bottom, you can simply make a new armature, but this time lay the coil flat on the table when you scrape the insulation off of the top half of the free wire ends.

A Bigger Motor

This next motor is simply a larger version of the first one, with a base made of wood.

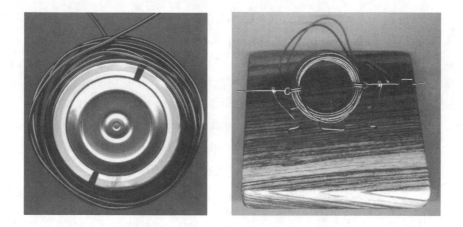

Place the magnet in the middle of the base. Drill four small holes around the magnet for the support wires.

Wind the coil using thick wire (20-gauge enameled copper wire is shown). Use a D cell as the coil form.

Use brass wire for the supports; make all the connections under the base so that everything looks nice and neat. For a battery connection, use a 9-volt battery clip.

WHY DOES IT DO THAT?

When electricity flows through a coil of wire, the coil becomes an electromagnet. An electromagnet acts just like a regular magnet: it has a north pole and a south pole, and can attract and repel other magnets.

The coil in the motor above becomes an electromagnet when the bare copper half of the armature wires touches the bare wire of the supports, and electricity flows into the coil. The electromagnet has a north pole that is attracted to the south pole of the regular magnet. It also has a south pole that is repelled by the south pole of the regular magnet.

When you scraped off the insulation from the armature wires, you did it with the coil standing up, and not lying flat on the table. This made the poles of the electromagnet point to the left and right, as if there were an invisible regular magnet that had the wire wrapped around it. If the coil were flat on the table, the poles would point up and down.

Since the poles point left and right, they have to move in order to line up with the magnet at the bottom, whose poles are aligned up and down. So the coil rotates to line up with the magnet. But once the coil is exactly lined up with the magnet, the insulated half of the wire touches the supports instead of the bare half. The electricity is cut off, and the coil is no longer an electromagnet. This leaves it free to coast on around until the bare copper can again touch the bare support and start the whole process over again.

A Faster Motor

One easy way to make the motor run faster is to add another magnet. Hold a magnet over the top of the motor while it is running. As you move the magnet closer to the spinning coil, one of two things will happen. Either the motor will stop or it will run faster. Which of these happens will depend on which pole of the magnet you have facing the coil. Make sure you hold the motor down so the magnets will not jump together and crush the motor!

There is another way to speed up the motor. The motor only gets electricity during half of its cycle. During the other half, the insulation blocks the flow of current. This is necessary because if you let the current continue to flow, the coil will spin around to face the magnet and will stop, facing the magnetic pole to which it is attracted.

But suppose instead of just stopping the current, you reversed it, so the north pole of the electromagnet became the south pole and vice versa. Midway through a rotation the coil would want to flip over again, and since it is already turning in that direction, it will continue to go in that directon due to the inertia and momentum of the coil. If you could get the current to reverse, and get it to happen at the right time, it would spin even faster.

The solution turns out to be pretty easy. Place the motor in front of you so that the axle goes left to right. Now attach a bare wire to the left support and let it rest on the right axle, just past the right support. Do the same thing with the right support and the left axle.

On one half of the cycle, the bare half of the axle will face down and touch the bare wire of the support, just like before. On the other half of the cycle, the bare half of the axle will touch the new wires that are resting on top of the axle. Since these wires are connected to the opposite supports, the current will flow in the opposite direction. The motor will get two kicks per cycle instead of one, and will never be coasting. It will always have power and will go twice as fast.

Look at the photo below of a motor built this way. The connections are hidden underneath the base for neatness, but you can see the wires resting on the top of the axles, and know that they are connected to the opposite supports.

What follows is a close-up view of the same motor. Notice that there are two tiny glass beads on the axles. These beads speed up the motor even more by reducing the friction of the armature against the supports. Since this reduction in friction balances the extra friction of the new wires, the motor still goes about twice as fast as the older, simpler motor.

A Motor with Two Coils

Can you construct a motor without any permanent magnets? Yes! In place of the magnet, use another coil of wire. This coil is called the *field coil*, while the coil that moves is called the *armature coil*.

The simplest way to build this motor would be to replace the permanent magnet with a coil of wire connected to a second battery. However, you can save the second battery, and waste less electricity, by arranging the coils as shown in the diagram at right:

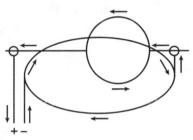

The diagram shows how the electrons flow through the coil. They start where the negative terminal of the battery is connected (marked here with a minus sign), then flow around and around inside the field coil until they come to the first support loop on the right. From there they flow into the armature coil, going around and around until they come out the other end, flowing into the second support loop on the left. From there they travel back to the positive terminal of the battery.

You can see that when the electricity flows through one coil, it also flows through the other. When the armature turns over, and the insulation cuts off the flow of electricity to the armature, it cuts off the flow of electricity to the field coil at the same time. Because both coils turn on and off together, there is never a time when one coil is on when the other is off— that would be a waste of electricity.

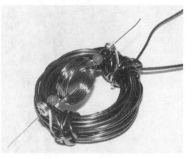

The photograph shows the completed motor. There are three different types of wire to more easily show how it goes together. The heavy wire is the field coil. It was wound with about 50 turns of wire around a D battery, then removed. The wire ends were wrapped around the coil to keep it tight, just as with the armatures of all the previous motors.

One end of the field coil wire is stripped of its insulation and formed into a loop for the first support. The other end goes to the battery.

A second wire forms the other support. One end is stripped and formed into a loop, and the other end winds around the coil a few times to make it sturdy, then goes to the other terminal of the battery.

The armature coil is made of thin wire and is formed in the same way as in the other motors. A couple of plastic beads keep the armature centered (they are optional). This motor works fine running from a D cell but runs even faster using a 9-volt battery.

A High-Voltage Motor

This device is very simple to build. You can put it together in five minutes using a few things found around the house. It's a high-voltage motor that acts like a bell, with a clapper that bangs furiously from one can to the other and back again, sometimes several

times per second. Occasionally, a big blue spark snaps between the cans, to add interest to the frenetic activity.

SHOPPING LIST

- ⊃ 2 empty soda cans
- ⊃ 5 inches of sewing thread
- ⊃ 2 square feet of aluminum foil
- ⊃ Cellophane tape
- ⊃ 2 wires (alligator test leads work great)
- ⊃ Plastic rod, such as an empty ballpoint pen

The photo below may be all you need to get this motor working.

Remove the pull-tabs from both cans, and discard one of the pull-tabs.

Tie the thread to the other pull-tab. Tape the other end of the thread to the center of the plastic rod.

Place the two cans side-by-side, 2 to 3 inches apart.

Balance the plastic rod on top of the two cans so that the pull-tab dangles freely about an inch from the table, midway between the cans.

Tape the bare end of one wire to the left can; this is the ground wire. The free end should be connected to an electrical ground, such as a cold water pipe. If a good electrical ground is not convenient, you can just hold onto the free end since your body is a good enough ground for this device.

Tape the other wire to the can on the right. Its free end will be connected to a source of high voltage. This is easier than it sounds, since a safe source of high voltage is right in front of you when you watch television or use a computer with a CRT monitor.

In the photo below, you can see that the device is sitting on top of a television. About 2 square feet of aluminum foil is pressed onto the face of the TV screen. It sticks there because the TV screen is highly charged with electricity. The free end of the right can's wire is attached to the aluminum foil.

To start the device, turn on the TV. The pull-tab is pulled to one can, but when it hits it, it gets pulled over to the other can, and then repeats.

᷑ WHY DOES IT DO THAT? ᷑

Inside a television, high voltage is used to send electrons to the screen at high speed, creating the picture. By placing a large conductor on the front of the screen, you can make a simple capacitor to tap into some of that high voltage and put it to use outside of the television. The voltage is high but the current is very small, so that touching the foil is no more harmful than touching a doorknob after scuffing your feet on the carpet.

In this experiment the can on the right is connected to the high voltage. The can on the left is connected to the ground, which can absorb all of the voltage it is sent and still be ready for more. The pull-tab and the can on the left start out without any electrical charge. We say they are at "ground potential."

When the can on the right is charged with a lot of free electrons from the foil on the TV screen, the electrons repel the electrons in the pull-tab and attract the positive nuclei in the pull-tab.

The electrons in the pull-tab move to the side farthest from the high voltage can on the right. This leaves the right side of the pull-tab more positive than the left side. The positive side of the pull-tab is attracted to the highly negative can on the right, and the pull-tab jumps over to touch the can on the right.

Once the pull-tab touches the can, the electrons from the can rush onto the pull-tab until it has the same high-voltage charge as the can it is touching. The pull-tab and the can now have the same charge, and like charges repel. The charged pull-tab is now repelled by the can on the right, and moves to the left.

The electrons in the can on the left are repelled by the pull-tab and move to the left side of the can, leaving the right side somewhat positive. This positive side attracts the negatively charged pull-tab and draws it up to touch the can.

Now the excess electrons on the pull-tab move onto the left can and into the ground. The pull-tab is now at ground potential again. It swings back toward the can on the right, and the whole process starts over again.

A Fancier Version

The device you have just built is called "Franklin's Bells," after a description of the device given by Benjamin Franklin. He used it to detect the approach of lightning storms. He connected one end to his lightning rod on top of the house, and the other end to an iron water pump well connected to the ground. Of course Ben didn't use soda cans; he used bells.

I built a pretty version with bells and set it up on top of the TV just like the version above. I left the clappers on the bells even though they no longer served any purpose. The bells were connected to the aluminum foil and the ground by thin copper wires you can barely see in the photo. The bells were purchased at a craft store, and the clapper is one of the small round bells called "jingle bells."

A Rotary High-Voltage Motor

At a company I worked for, we had a contest: The Most Creative Use of Office Supplies. The device below would clearly be a contender. It is based on a wonderful design by Bill Beaty, but this version uses only objects found around the typical office.

Using the safe high-voltage power you get by placing a sheet of aluminum foil on the face of a television or computer CRT screen, it spins a styrofoam cup around at a respectable speed.

SHOPPING LIST

- 2 empty soda cans
- Styrofoam cup (a paper cup will also work)
- 2 ballpoint pens (the simple nonclicking type)
- 2 square feet of aluminum foil
- 2 paper clips
- Cellophane tape
- 2 wires (alligator test leads work great)
- Paper plate

TOOLS

- Sharp knife
- Hot glue gun (or regular glue if you don't mind waiting)

Start by spreading a thin layer of glue over the outside of a styrofoam cup. Before the glue dries, cover the cup with aluminum foil. Press the foil flat against the cup to flatten any wrinkles.

With a sharp knife, neatly cut a half-inch strip out of the foil on both sides so you have two patches of foil, one on each side of the cup, that do not touch one another.

The cup will be spinning upside down on the tip of a ballpoint pen. To keep the cup centered on the pen point, and to provide a low friction bearing, you must glue something hard to the center of the bottom of the cup, something that has a little dimple in it to sit on the pen point. I chose to sacrifice the end of another ballpoint pen.

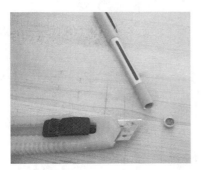

The photo at left shows the end of the pen, cut off with a sharp knife. The side of the cut end that is facing down has a little dimple that is perfect as a place to accept the point of the other ballpoint pen.

Glue the end of the pen in the exact center of the inside bottom of the cup, as shown below. Note the little dimple.

Next, make the stand for the motor. Start with a paper plate, face down, and glue the bottom of a ballpoint pen to the exact center of the plate so the pen stands vertically.

Glue two soda cans upside down onto the plate, leaving enough room between them for the styrofoam cup to rotate easily without touching either can. There should be about a half-inch gap between the cup and either can.

Straighten two of the bends of a paper clip, leaving one end bent as shown below, and tape them to the cans. Bend the wires

into an S shape, leaving enough room to place the cup on top of the pen.

Now put the foiled cup upside down onto the pen point. Make sure the dimple fits onto the pen point. The wires should be about a half-inch away from the cup, with the point being closest to the cup. Nothing should be touching the cup except the point of the pen.

Connect a wire from the can on the right to a large sheet of aluminum foil pressed against the screen of a TV (or a computer with a CRT screen). Connect another wire to the left can, and connect the free end to a good ground

connection, such as a cold water pipe or the metal frame of a computer. In a pinch, you can just hold onto the free end since your body is a good enough ground for this little motor.

When you turn on the television, the foil will pick up a high voltage and the motor will start spinning. As it slows down, turn the television off; the motor will get another

kick, and spin faster. You can keep this up as long as you feel like turning the TV on and off.

🤖 WHY DOES IT DO THAT? 🤖

On this device, the can on the right is charged with high voltage from the face of the TV. This means that the electrons are pushed onto the can with a lot of electrical "pressure." The electrons all have the same negative charge, so they repel one another.

The electrons are most crowded at the point of the wire, creating the highest pressure. Pressure is so high that electrons can get pushed right off the wire onto the air molecules near the wire. These air molecules thus become negatively charged, causing the extra electrons in the wire to repel them. They move away from the wire and hit the aluminum foil, where the electrons leave the air molecules and collect on the foil.

Meanwhile, the can on the right, which still has an excess of electrons, repels the electrons on the foil. The electrons in the foil repel the electrons in the can on the left, so they move away from it, leaving behind the positively charged nucleus, which attracts the electrons in the foil.

With the right can repelling and the left can attracting, the foil is pulled to the left, and the cup rotates. Half a revolution later, the charged foil is next to the wire that is attached to the ground. The electrons from the foil charge the air around the wire, and the electrons can move from the foil, through the air and the wire, to the ground. This leaves the foil uncharged so the can on the right does not repel it as it moves by. At this point, the foiled cup is back to where it started, and the process starts over again.

A Simple Homemade Van de Graaff Generator

In the previous two projects, the devices stole high voltage from a television set to power the high-voltage motors. In this project, you will build a device that can generate 12,000 volts from an empty soda can and a rubber band.

The device is called a Van de Graaff Generator. Science museums and research facilities have large versions that generate potentials in the hundreds of thousands of volts. This device is more modest but is still capable of drawing ½-inch sparks from the soda can to a waiting finger. The spark is harmless and similar to the jolt you get from a doorknob after scuffing your feet on the carpet.

SHOPPING LIST

- 5 x 20-mm GMA-type electrical fuse (RS #270-1062)
- Small DC motor (such as RS #273-223)
- Battery clip (RS #270-324)
- Battery holder (RS #270-382)
- 2 6-inch-long stranded electrical wires (such as from an extension cord)
- 2 6-inch pieces of ¾-inch PVC plumbing pipe
- ¾-inch PVC union coupler
- ¾-inch PVC "T" connector
- Empty soda can
- Small nail
- Rubber band, ¼ inch by 3–4 inches
- Styrofoam cup (a paper cup will also work)
- Block of wood
- Small glass tube
- Electrical tape
- Plastic rod

TOOLS

- Hot glue gun (or regular glue if you don't mind waiting)
- Drill
- Sharp knife
- Saw or PVC cutter

The shopping list sounds like a lot of stuff, but you will find that the whole project can easily be put together in an evening following the step-by-step photos below, once all the parts have been collected.

Start at the bottom and work your way up. First, cut a 2- to 3-inch-long piece of ¾-inch PVC pipe with a saw or PVC cutter and glue that to the wooden base. This piece will hold up the generator and allow it to be removed more easily, in order to replace the rubber band or make adjustments.

The PVC "T" connector will hold the small motor. The motor fits too loosely by itself, so you must wrap paper or tape around it to make a snug fit. The shaft of the motor can be left bare, but the generator will work a little better if it is made fatter by wrapping tape around it or, even better, by putting a plastic rod with a hole in the center onto the shaft to act as a pulley for the rubber band.

Next, drill a small hole in the side of the PVC "T" connector, just under the makeshift pulley on the motor. This hole will be used to hold the lower "brush," which is simply a bit of stranded wire frayed at the end that almost touches the rubber band on the pulley.

As the photo shows, the stranded wire is held in place with electrical tape or some other tape or glue. The rubber band is now placed around the pulley and allowed to hang out the top of the "T" connector.

Next, cut another 2- or 3-inch piece of ¾-inch PVC plumbing pipe. This will go into the top of the "T" connector, with the rubber band going up through it. Use the small nail to hold the rubber band in place, as shown below. The length of the PVC pipe should be just long enough to fit the rubber band. The rubber band should not be stretched too tightly since the resulting friction would prevent the motor from turning properly and increase wear on the parts.

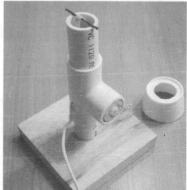

Cut a styrofoam cup about an inch from the bottom, and carefully cut a ¾-inch-diameter hole in the center of the bottom of the cup. This hole should fit snugly onto the ¾-inch PVC pipe.

Now drill three holes near the top of the PVC union coupling. Two of these holes need to be directly opposite one another since they will hold the small nail, which will act as an axle for the rubber band. The third hole should be drilled between the other two, and it will hold the top "brush," which, like the bottom brush, will almost touch the rubber band.

Tape the top brush to the PVC union coupler, and place the coupler on the ¾-inch pipe above the styrofoam cup collar. Thread the rubber band through the coupler and hold it in place with the small nail, as before.

Bare the top brush so it has no insulation, and twist it to keep the individual wires from coming apart. You can solder the free end if you like, but it is not necessary. The free end of the top brush will be curled up inside the empty soda can when you are done, and thus electrically connect the soda can to the top brush.

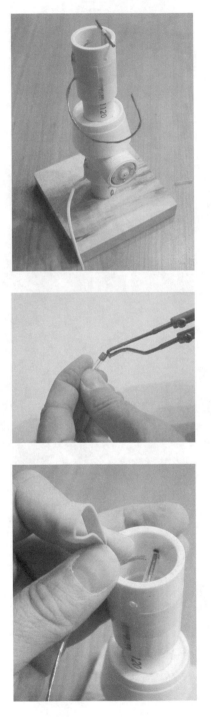

A small glass tube will act as both a low-friction top pulley and a triboelectric complement to the rubber band, to generate static electricity by rubbing. Glass is one of the best materials to rub against rubber to create electricity. You can make a tube by taking apart a small electrical fuse. The metal ends of the fuse come off easily if heated with a soldering iron or match. The solder inside the fuse drips out when the ends come off, so be careful. The glass, the metal cap, and the molten solder are all quite hot and will blister your skin if you touch them before they cool.

Save the metal caps—you will use them in a future project.

The resulting glass tube has nice, straight, even edges, which are "fire polished" for you, so there is no sharp glass and no uneven edges to catch on the PVC and break the glass.

The next step is a little tricky. Place the small nail through one of the two holes in the PVC union coupler, and place the small glass tube on the nail. Then place the rubber band on the glass tube, and the nail is in the second hole. The rubber band is on the glass tube, which is free to rotate around the nail.

Now glue the styrofoam collar in place on the PVC pipe. At this point you are ready for the empty soda can. Aluminum pop-top cans are good for high voltage because they have nice, rounded edges, which minimize "corona discharge" (see page 60).

With a sharp knife, carefully cut out the top of the soda can. Leave the nice crimped edge and cut close to the side of the can to avoid sharp edges. You can smooth the cut edge by "stirring" the can with a metal tool like a screwdriver, pressing outward as you stir.

Tuck the free end of the top brush wire into the can and invert the can over the top of the device until it rests snugly on the styrofoam collar.

The last step is to attach the batteries. I like to solder a battery clip to the motor terminals and then clip this onto either a 9-volt battery or a battery holder for two AA batteries. The 9-volt battery works but it runs the motor too fast, making a lot of noise, and risks breaking the glass tube. It does, however, create a slightly higher voltage ... until the device breaks.

To use the Van de Graaff Generator, simply clip the battery to the battery clip. If the brushes are very close to the ends of the rubber band, but not touching, you should be able to feel a spark from the soda can if you bring your finger close enough. It helps to hold onto the free end of the bottom brush with the other hand while doing this.

If you want, you can use the generator to power the Franklin's Bells from the previous project. Just clip the bottom brush

wire to one bell and attach a wire to the top of the generator, connecting it to the other bell. The pop-top clapper of the Franklin's Bells should start jumping between the soda cans. It may need a little push to get started.

WHY DOES IT DO THAT?

You may have at one time rubbed a balloon on your hair and then made the balloon stick to the wall. If you have never done this, try it! The Van de Graaff Generator uses this phenomenon and two others to generate the high voltage needed to make a spark.

The First Phenemonon

When a balloon makes contact with your hair, the molecules of the rubber touch the molecules of your hair. When they touch, the rubber's molecules attract electrons from the hair's molecules. When you take the balloon away, some of those electrons stay with the balloon, giving it a negative charge.

When the balloon is held near a wall, the extra electrons on the balloon repel the electrons in the wall, pushing them back from the surface. The surface of the wall is left with a positive charge since there are fewer electrons than when it was neutral. The positive wall attracts the negative balloon with enough force to keep it stuck to the wall. If you collected a bunch of different materials and touched them to one another, you could find out which ones were left negatively charged and which were left positively charged.

You could then take these pairs of objects and put them in order in a list, from the most positive to the most negative. Such a list is called a Triboelectric Series. The prefix tribo- means "to rub."

THE TRIBOELECTRIC SERIES

Most positive (items at this end lose electrons)

Asbestos
Rabbit fur
Glass
Hair
Nylon
Wool
Silk
Paper
Cotton
Hard rubber
Synthetic rubber
Polyester
Styrofoam
Orlon
Saran
Polyurethane
Polyethylene
Polypropylene
Polyvinyl chloride (PVC pipe)
Teflon
Silicone rubber

Most negative (items at this end steal electrons)

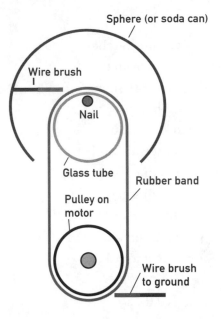

Sphere (or soda can)

Wire brush

Nail

Glass tube

Rubber band

Pulley on motor

Wire brush to ground

The Van de Graaff Generator you built uses a glass tube and a rubber band. The rubber band steals electrons from the glass tube, leaving the glass positively charged and the rubber band negatively charged.

The Second Phenomenon

Triboelectric charging is this device's first trick. The second involves the wire brushes.

When a metal object is brought near a charged object, something quite interesting happens. The charged object causes the electrons in the metal to move. If the object is charged negatively, it pushes the electrons away. If it is charged positively, it pulls the electrons toward it.

Electrons are all negatively charged. Because like charges repel, and electrons are all the same charge, electrons will always try to get as far away from other electrons as possible. If a metal object has a sharp point on it, the electrons on the point are pushed by all of the other electrons in the rest of the object. So on a point, there are a lot of electrons pushing from the metal but no electrons pushing from the air.

If there are enough extra electrons on the metal, they can push some electrons off the point and into the air. The electrons land on the air molecules, making them negatively charged. The negatively charged air is repelled from the negatively charged metal, and a small wind of charged air blows away from the metal. This is called "corona discharge" because the dim light it gives off looks like a crown.

The same thing happens in reverse if the metal has too few electrons (if it is positively charged). At the point, all of the positive charges in the metal pull all the electrons from the point, leaving it very highly charged.

The air molecules that hit the metal point lose their electrons to the strong pull from the positive tip of the sharp point. The air molecules are now positive, and are thus repelled from the positive metal.

The Third Phenomenon

There is one more phenomenon the Van de Graaff Generator uses.

Remember that all electrons have the same charge, so they all try to get as far from one another as possible. The third phenomenon uses the soda can to take advantage of this feature of the electrons in an interesting way. If the soda can is given a charge of electrons, they will all try to get as far away from one another as possible. This makes all the electrons crowd to the outside surface of the can. Any electron on the inside of the can will feel the push from all the other electrons and will move. The electrons on the outside feel the push from the can, but they do not feel any push from the air around the can, which is not charged. This means that you can put electrons on the inside of the can and they will be pulled away to the outside. Electrons can continue to be added to the inside of the can, and they will always be pulled to the outside.

Putting All Three Phenomena Together

So now look at the Van de Graaff Generator with the three phenomena in mind.

The motor moves the rubber band around and around. The rubber band loops over the glass tube and steals the electrons from the glass. The rubber band is much bigger than the glass tube. The electrons stolen from the glass are distributed across the whole rubber band.

The glass, on the other hand, is small. The negative charges that are spread out over the rubber band are weak compared to the positive charges that are all concentrated on the little glass tube.

The strong positive charge on the glass attracts the electrons in the wire on the top brush. These electrons spray from the sharp points in the brush, and charge the air. The air is repelled from the wire and attracted to the glass.

But the charged air can't get to the glass because the rubber band is in the way. The charged air molecules hit the rubber and transfer the electrons to it.

The rubber band travels down to the bottom brush. The electrons in the rubber push on the electrons in the wire of the bottom brush. The electrons are pushed out of the wire and onto whatever large object we have attached to the end of the wire, such as the Earth or a person.

The sharp points of the bottom brush are now positive, and they pull the electrons off any air molecules that touch them. These positively charged air molecules repel the positively charged wire and attract the electrons on the rubber band. When they hit the rubber, they get their electrons back, and the rubber and the air both lose their charge.

The rubber band is now ready to go back up and steal more electrons from the glass tube. The top brush is connected to the inside of the soda can. It is positively charged, and so attracts electrons from the can. The positive charges in the can move away from one another. (They are the same charge, so they repel, just like electrons.) The positive charges collect on the outside of the can, leaving the neutral atoms of the can on the inside, where they are always ready to donate more electrons.

The combined effect is to transfer electrons from the soda can into the ground, using the rubber band like a conveyor belt. It doesn't take very long for the soda can to lose so many electrons that it becomes 12,000 volts more positive than the ground.

When the can gets a very positive charge, it eventually has enough charge to steal electrons from the air molecules that hit the can. This happens most at any sharp points on the can. If the can were a perfect sphere, it would be able to reach a higher voltage since there would be no place where the charge was more concentrated than anywhere else.

If the sphere were larger, an even higher voltage could be reached before it started stealing electrons from the air, because a larger sphere is not as "sharp" as a smaller one. The places on our soda can where the curves are the sharpest are where the charge accumulates the most, and where the electrons are stolen from the air.

Air ionizes in an electric field of about 25,000 volts per inch. Ionized air conducts electricity like a wire does. You can see the ionized air conducting electricity because it gets so hot it emits light—what we call a spark. Since our generator can draw sparks that are about a half-inch long, we know we are generating about 12,500 volts.

Try This

One of the fun things to do with a Van de Graaff Generator is to show how like charges repel.

Take a paper napkin and cut thin strips of the lightweight paper. Then tape the ends of the paper together at one end, and tape that end onto the Van de Graaff Generator. The effect will look somewhat like long hair cascading down the soda can.

Now turn on the generator. As the thin strips of paper receive the same charge, they start to repel from one another. The effect is hair-raising. The strips start to stand out straight from the can, like the hair on a scared cat's back.

A High-Voltage Ion Motor

This motor is very simple to build, taking only a few minutes. All you need is two pieces of wire, the small metal cap left over from the fuse you took apart in the previous project, scissors, and some cellophane tape. The motor creates an ion "wind" that spins it around like a helicopter.

First, take one piece of wire—a straightened paper clip will do—and cut the end at an angle so it is sharp. Bend the other end into a rough loop or triangle, so the wire will stand up with the sharp point facing straight up. A little tape will help hold it onto a table or a block of wood.

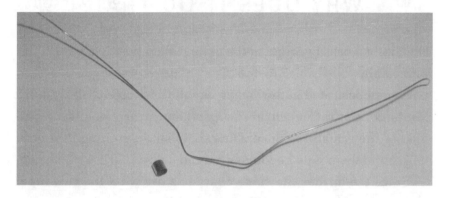

The armature (the part that spins) is made from the other piece of wire and the metal fuse cap. Sharpen both ends of the wire by cutting the ends at a diagonal, like you did with the base wire. Bend the wire into an S shape. The pointed ends of the wire should point at 90 degrees from the center straight part of the wire.

Attach the metal cap to the center of the wire with tape. Place the cap onto the pointed end of the base wire, and bend the S-shaped ends of the armature wire down, so it will easily balance on the sharp end of the base wire. The armature should now spin freely if you tap it gently.

Connect a source of high voltage to the base wire using an alligator clip or a wire. The high-voltage source can be the Van de Graaff Generator or just a few square feet of aluminum foil pressed against the front of your television set. As the high voltage is turned on, the armature will start to spin in the direction away from the sharp points. The Van de Graaff Generator may need a good ground, or a person holding onto the ground wire.

The television will give the motor a good kick every time it is turned on or off, and turning it on and off every second will get the motor spinning quite rapidly.

🤖 WHY DOES IT DO THAT? 🤖

The motor works by ionizing the air and then pushing against the ionized air.

As demonstrated in the Van de Graaff Generator project, sharp points concentrate electric charges. The sharp points on the ends of the armature concentrate the charges so much that the air around the points becomes charged as well.

Since the air has the same charge as the wire, the two repel one another. You can actually feel a small wind coming from the sharp point. As the wire pushes on the charged air, they both move away from one another. The air blows away, and the wire spins.

3

ELECTROCHEMISTRY

The Plastic Hydrogen Bomb

It sounds a bit dangerous, doesn't it? But this device actually demonstrates the principles of electrochemistry. It's also a high-tech squirt gun.

The Plastic Hydrogen Bomb uses electricity to break water molecules into hydrogen and oxygen. It then uses a spark of electricity to explosively recombine the gases into high-pressure steam, which propels a stream of water high into the air.

Its construction is a little more difficult than the other devices in this book, but the skills you will learn by building it can be put to good use building many other devices and works of art.

SHOPPING LIST
- Polyester resin and catalyst
- 2 carbon rods from cheap batteries (or large gold plated connectors)
- Piezoelectric igniter from a Scripto electronic lighter

- ➲ 9-volt battery clip
- ➲ Paraffin wax (from a cheap white candle)
- ➲ Insulated copper wire (about 20–22 gauge)
- ➲ Solder
- ➲ Clear plastic box (an empty Tic-Tacs box works well)

TOOLS

- ☐ Soldering gun
- ☐ Pencil
- ☐ Scissors
- ☐ Tupperware bowl
- ☐ Drill
- ☐ ¼-inch drill bit

How the Bomb Works

In schematic form, the bomb looks like this:

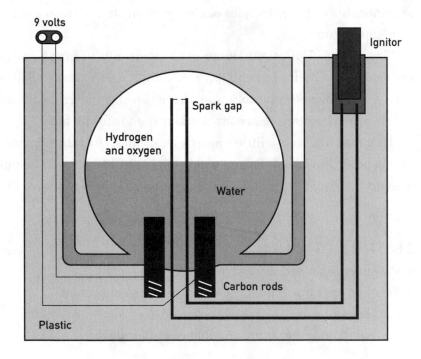

A 9-volt battery is connected to two carbon rods that are under water in the bomb chamber. Electrical energy is converted into chemical energy as the water breaks up into hydrogen and oxygen, which is trapped in the upper half of the chamber. When a piezoelectric igniter from the electronic lighter is pushed, it causes a spark to jump the spark gap, igniting the hydrogen and oxygen. The resulting high-pressure steam forces the water out through the exit tubes, high into the air.

My First Few Attempts at Constructing a Hydrogen Bomb

The first hydrogen bomb I built was done entirely with the "lost-wax casting" method used by artists, sculptors, and jewelers. In the lost-wax method, a model of the finished object is made of wax, then plaster is poured over the wax and allowed to harden. The wax is then melted and poured out of the hardened plaster. The resulting hollow cavity is filled with molten metal, which cools into the same shape as the original wax model. The plaster is broken away, and the finished piece of art or jewelry is cleaned and polished.

With my bomb, I eliminated all of the steps after the wax had been melted out of the mold, since my objective was to build a hollow in the plastic into which the carbon rods and spark gap protrude.

My first bomb is shown below:

Note that the spark gap at the top of the bomb is made out of carbon rods. Later versions use a simpler spark gap made of copper wire.

You can see the red wire from the battery clip going to the carbon rod at the bottom of the hemispherical chamber. You can also see a rectangular tube curving up from the bottom of the chamber and exiting at the top of the device. This tube, formed from a thin sheet of wax, was there to let the water in and out of the chamber. There was another tube like it on the other side. The wax model sat on a pile of plastic beads, and the liquid plastic was poured over it.

The problem with the lost-wax version was that you couldn't see the inner workings very well. My next attempt used a clear plastic box and some clear plastic tubing. It exploded in my face on my first attempt to use it. The remains are shown at left.

The plastic box was not strong enough to contain the force of the explosion, and the plastic tubes were too narrow and too long to let the water out fast enough to prevent the plastic from being blown apart. However, the new spark gap, made of twisted copper wire, worked perfectly.

In succeeding refinements, the plastic box was completely covered with the polyester resin, so that the walls of the chamber were at least ½ inch thick. The entry/exit tubes were drilled into the plastic so they were straight and wide. A little bit of lost-wax technology was used to provide a target for the drill, and so the wax could support the carbon rods and spark gap.

Constructing Your Bomb

Construction starts with the removal of two carbon rods from a couple of old-fashioned carbon-zinc batteries. These are the cheap type, such as the Eveready "Classic" or Radio Shack "Heavy Duty" types, not the alkaline batteries that have largely replaced them in common use.

In the photo on the next page, one of the batteries has the

cover and top removed to show the carbon rod sticking up out of the cardboard seal. A little twist and pull and the rod slips right out, and can be washed and used.

Next, attach insulated copper wire to the carbon rods by stripping the insulation off of a few inches of the wire on one end, and winding it several turns around the rod as shown, to make a good connection. The wire is tightly twisted to hold the connection securely.

Bend the insulated copper wire around a pencil, then twist the ends tightly for a couple of inches to make the spark gap. Then cut the loop and trim some of the insulation from the ends.

Place the spark gap and the carbon rods in the clear plastic box. (The box from Tic-Tacs candies will do if you don't have a cubical box like the one shown in the photo.) Fill the box with water within about a half-inch from the top. Then pour melted wax on top of the water to seal the box and the spark gap and rods. After the wax has hardened, loosen it a little to let the water out, then replace it to seal the box.

Now attach two thick pieces of wax to the bottom that will form hollow channels to let the water in and out. If you like, you can extend these all the way up to the top, making channels of square cross-section. I like to drill the channels later, but you might

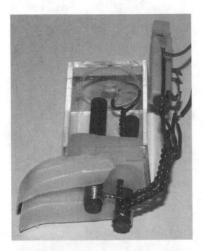

like challenging people to guess how you drilled square holes in the plastic.

In the next step, take apart an electronic lighter to get the piezoelectric igniter. (You can find larger igniters, which are easier to solder, in lighters used for fireplaces.)

The igniter is the little gadget that produces a spark when it is pushed down. It has two contacts that need to be roughened with a knife or sandpaper before the spark gap's wires can be soldered to them.

You can see how the larger igniter is soldered to the spark gap in the photo of the box on its side in the top photo. The wires to the 9-volt battery clip are also soldered to the wires that go to the carbon rods.

The casting of the polyester resin is done in three steps. Use a Tupperware bowl for the mold. First cast a thin layer on the bottom of the bowl. Allow this layer to harden; it will be the base on which the carbon rods and the rest of the apparatus will sit.

Next, cast a half-inch-thick layer to hold the carbon rods in place, and cover the wax outriggers at the bottom of the box. Allow this layer to harden before pouring the last layer. This will

ensure that the hollow box will not float to the top of the mold when the last layer is poured.

Finally, pour the last layer, completely covering the top of the box to a thickness of at least a half-inch. It should also cover the soldered connections on the piezoelectric igniter, but must not touch the moving parts or get inside the igniter, or the igniter will become stuck and will not operate.

The penultimate step is drilling the entry and exit holes. I use a ¼-inch drill bit. Finally, melt the wax out by placing the project in a warm oven, no more than 150°F, on top of a newspaper-lined pan to catch the melted wax.

A Slight Variation

This project uses carbon rods because they will not disintegrate when electric current is run through the water. If you used copper wire, one wire would be eaten away and the other would get a plating of copper sludge.

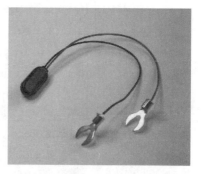

However, if you use a less reactive metal, such as gold, this wire destruction will take place much more slowly, and the device can be used many times without any noticeable changes.

To test this, I made a bomb similar to the one we just made, but with gold-plated connectors used for highend stereo connections.

The result is a nice looking Hydrogen Bomb.

Operating the Bomb

To operate the bomb, hold it under a faucet, tilted so that water can enter one hole and air can exit the other. The chamber should be only about half full, with a bubble of air keeping the spark gap dry.

Plug in a 9-volt battery and watch as tiny bubbles of hydrogen and oxygen form on the carbon rods (or gold electrodes).

After about 15 to 20 minutes, enough of the gases will have formed for the bomb to explode with a nice effect. Note that as the gases form, the water is displaced and leaks out the drilled holes. Hold the bomb at arm's length and press the igniter down until it clicks. Be very careful—the water remaining in the bomb and the tubes will shoot out, covering the ceiling and spectators.

Several years ago I was demonstrating my science toys on a television show. Before the show began taping, we filled a bomb and plugged in the battery. After describing and demonstrating other projects for about 25 minutes, it was almost time to go. I held the bomb in my hands and clicked on the igniter, but nothing happened. I clicked a few more times, muttering something about the spark gap being wet.

The host of the show decided to give it a try. He held the bomb and carefully looked down the tubes as he clicked the igniter. Of course it worked for him. The water splashed up into his glasses, his hair, his shirt, the ceiling, all over. The camera crew was laughing so hard you could see the camera shaking.

Always be careful with the Plastic Hydrogen Bomb.

WHY DOES IT DO THAT?

Using electricity to break up water is called *electrolysis*, Greek for "loosening by electricity."

Water is made up of two atoms of hydrogen and one atom of oxygen: H_2O. In the liquid form, the molecules are constantly breaking up into electrically charged pieces and then getting back together. The electrically charged pieces are called ions.

Water breaks up into ions by losing one of the hydrogen atoms. The nucleus of the hydrogen atom breaks away, leaving its electron behind with the other two atoms, creating a positively charged hydrogen ion, H^+, and a negatively charged hydroxyl ion, OH^-.

On your bomb, one of the carbon rods is attached to the negative terminal of the battery and has excess electrons. This electrode is called the *cathode*. The excess electrons attract the positively charged hydrogen atoms to the cathode.

Two of the electrons on the cathode combine with two H^+ ions, forming a hydrogen molecule, H_2. This molecule joins others and forms bubbles of hydrogen gas that rise through the water, making room for more water to contact the cathode and form more hydrogen.

The reaction at the cathode is described chemically as:

$$2H^+ + 2e^- \longrightarrow H_2$$

The other electrode is called the *anode*. The battery pulls electrons from the anode, leaving it with a positive charge. This positive charge attracts the negatively charged hydroxyl ions to the anode. Four hydroxyl ions get together and form two molecules of water and one molecule of oxygen, while donating four electrons to the anode.

The reaction at the anode is described chemically as:

$$4OH^- \longrightarrow 4e^- + 2H_2O + O_2$$

The oxygen (O_2) forms bubbles and rises through the water to join the hydrogen.

It takes a lot of energy to separate the H^+ ions from the OH^- ions. This energy is stored in the form of the hydrogen and oxygen gases. Some of the energy comes from the chemical bond between the two hydrogen atoms in the hydrogen molecule. Likewise, energy comes from the bond between the two oxygen atoms in the oxygen molecule. Most of the energy comes from the battery. All of this energy is "stored" by the mere separation of the gases into their respective molecules. If you could cause the atoms to rearrange themselves to form water again, that energy would be released.

In order to cause the gases to reform as water, you must first add a little energy to break the bond between the hydrogen atoms, and a little more to break the bonds between the oxygen atoms. Only when these bonds are broken will the atoms be free to rearrange to form other molecules. This energy is supplied by the spark. It breaks up the molecules of gas into their atoms, so they can recombine into H_2O, releasing the stored energy as heat.

The heat from those first recombinations is enough to break up more of the gas molecules, allowing them also to recombine into water. This reaction happens rapidly, consuming all of the gases and producing quite a bit of heat. Almost all of the energy that the battery put into the system in 20 minutes is released in a small fraction of a second.

The H_2O that results from the reaction is too hot to be a liquid. This gas's heat causes it to expand to take up much more space than the original gases needed. As it expands, it pushes the remaining water out the only exit that exists—the two holes pointing up toward the ceiling.

A Solar Cell You Can Make in Your Kitchen

A solar cell is a device for converting energy from the sun into electricity. The high-efficiency solar cells you can buy at Radio Shack and other stores are made from highly processed silicon and require huge factories, high temperatures, vacuum equipment, and lots of money.

If you are willing to sacrifice efficiency for the ability to make your own solar cells in the kitchen out of materials from the neighborhood hardware store, you can demonstrate a working solar cell in about an hour.

This solar cell is made from cuprous oxide instead of silicon. Cuprous oxide was one of the first materials found to display the photoelectric effect, in which light causes electricity to flow in a material.

Thinking about how to explain the photoelectric effect is what led Albert Einstein to the theory of relativity and to the Nobel Prize for physics.

SHOPPING LIST

- Sheet of copper flashing from the hardware store (about ½ square foot)
- 2 alligator clip leads
- Sensitive microammeter that can read currents between 10 and 50 microamperes (Radio Shack sells small LCD multimeters that will do, but I used a small surplus meter with a needle.)
- Electric stove (My kitchen stove is gas, so I bought a small one-burner electric hotplate. The little 700-watt burners probably won't work—mine is 1,100 watts, so the burner gets red hot.)
- Large, clear plastic bottle off of which you can cut the top, like a 2-liter spring water bottle (a large-mouth glass jar will also work)
- Table salt
- Tap water

TOOLS

- Sand paper or wire brush on an electric drill
- Sheet metal shears for cutting the copper flashing

Building the Cell

First, cut a piece of the copper flashing that is about the size of the burner on the stove. Wash your hands so they don't have any grease or oil on them. Then wash the copper piece with soap or cleanser to get any oil or grease off of it. Use sandpaper or a wire

 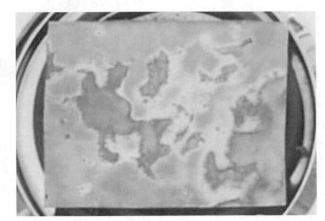

brush to thoroughly clean the copper piece to remove any sulfide or other light corrosion.

Next, place the cleaned and dried copper piece on the burner and turn the burner to its highest setting.

As the copper starts to heat up, beautiful oxidation patterns begin to form. Oranges, purples, and reds will cover the copper.

As the copper gets hotter, the colors are replaced with a black coating of cupric oxide. This is not the oxide we want, but it will flake off later, showing the reds, oranges, pinks, and purples of the cuprous oxide layer underneath. The last bits of color will disappear as the burner starts to glow red.

When the burner is glowing red hot, the sheet of copper will be coated with a black coat of cupric oxide. Let it cook for a half an hour so the black coating will be thick. This is important; a thick coating will flake off nicely, while a thin coat will remain stuck to the copper.

After the half-hour of cooking, turn off the burner. Leave the hot copper on the burner to cool slowly. If it cools too quickly, the black oxide will stay stuck to the copper.

As the copper cools, it shrinks. The black cupric oxide also shrinks. But because they shrink at different rates, the black cupric oxide will flake off. The little black flakes pop off the copper with enough force to fly a few inches. This will create a slight mess around the stove, but it is fun to watch.

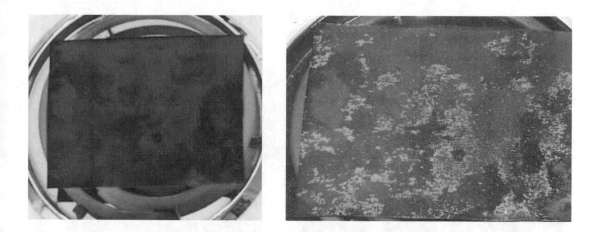

When the copper has cooled to room temperature—after about 20 minutes—most of the black oxide will be gone. Lightly scrub the sheet with your hands under running water to remove most of the small bits, but resist the temptation to remove all of the black spots by hard scrubbing or by flexing the soft copper. This might damage the delicate red cuprous oxide layer needed to make to solar cell work.

The rest of the assembly is very simple and quick.

Cut another piece of copper flashing about the same size as the first one. Bend both pieces gently so they will fit into the plastic bottle or jar without touching one another. The cuprous oxide coating that was facing up on the burner is usually the best side to face outward in the jar, since it has the smoothest, cleanest surface.

Attach two alligator clip leads, one to the new copper plate and one to the cuprous oxide coated plate. Connect the lead from the clean copper plate to the positive terminal of the microammeter. Connect the lead from the cuprous oxide plate to the negative terminal of the meter.

Now mix about 2 tablespoons of salt into some hot tap water. Stir the mixture until all the salt is dissolved. Then pour the saltwater into the jar, being careful not to get the clip leads wet. The saltwater should not completely cover the plates; you should leave about an inch of plate above the water so that you can move the solar cell around without getting the clip leads wet.

The photo at left above shows the solar cell in my shadow as I took a picture. Notice that the meter is reading about 6 microamps of current. Because the solar cell is a battery, even in the dark it will usually show a few microamps of current.

The photo at right shows the solar cell in the sunshine. Notice that the meter has jumped up to about 33 microamps of current. Sometimes it will go over 50 microamps, its needle swinging all the way over to the right.

A Note About Power

This solar cell produces 50 microamps at 0.25 volts. This is 0.0000125 watts (12.5 microwatts). Don't expect to power light bulbs or charge batteries with this device. It can be used as a light detector or light meter, but it would take acres of them to power your house.

🤖 WHY DOES IT DO THAT? 🤖

Cuprous oxide is a type of material called a *semiconductor*. A semiconductor is in between a conductor, where electricity can flow freely, and an insulator, where electrons are bound tightly to their atoms and do not flow freely.

In a semiconductor there is a gap, called a *bandgap*, between the electrons that are bound tightly to the atom and the electrons that are farther from the atom, which can move freely and conduct electricity. Electrons cannot stay inside the bandgap. An electron cannot gain just a little bit of energy and move away from the atom's nucleus into the bandgap. An electron must gain enough energy to move farther away from the nucleus, outside of the bandgap.

Similarly, an electron outside the bandgap cannot lose a little bit of energy and fall just a little bit closer to the nucleus. It must lose enough energy to fall past the bandgap into the area where electrons are allowed.

When sunlight hits the electrons in the cuprous oxide, some of the electrons gain enough energy from the sunlight to jump past the bandgap and become free to conduct electricity. The free electrons move into the saltwater, then into the clean copper plate, into the wire, through the meter, and back to the cuprous oxide plate. As the electrons move through the meter, they perform the work needed to move the needle. When a shadow falls on the solar cell, fewer electrons move through the meter, and the needle dips back down.

A Flat-Panel Solar Cell

You can make a more portable version of the solar cell in a flat panel form, using the clear plastic top from a plastic CD jewel case as the window, and lots of silicone rubber glue to both attach the pieces together and insulate them from each other.

First, make a cuprous oxide plate like you did in the first solar cell. Sand one corner clean, all the way down to the shiny copper, and solder an insulated copper wire to it for the negative lead.

Cut a U-shaped piece from the copper sheeting to create the positive plate, a little bit larger than the cuprous oxide plate, with the cutout portion of the U a little bit smaller than the cuprous oxide plate. Solder another insulated copper wire to one corner of the U, as shown middle left.

Glue the U-shaped copper plate to the plastic window. Use plenty of silicone glue to keep the saltwater from leaking out. Make sure that the solder connection is either completely covered with glue or is outside of the glue U. (Completely covering it in glue is best.)

The photo at bottom left shows the backside of the solar cell (the side not facing the sun) at this point in the construction.

The photo at bottom right shows the front side of the solar cell (the side that will face the sun). Notice that the silicone glue does not completely cover the copper, since

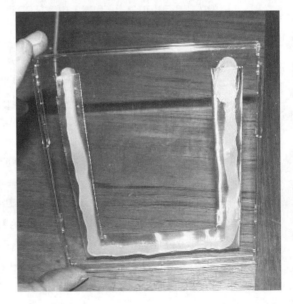

some of the copper must eventually be in contact with the saltwater.

Lay a good size bead of glue onto the clean U-shaped copper plate. This layer will act as an insulator between the clean copper plate and the cuprous oxide plate, and must be thick enough to leave some room for the saltwater. Again, not all of the copper should be covered; allow plenty of copper to contact with the saltwater.

Gently press the cuprous oxide plate onto this layer of glue. You should press hard enough to make sure the glue seals off any gaps, but not so hard that the two plates touch. The photo at left below shows the backside of the solar cell (the side not facing the sun) at this point.

The front side of the solar cell is shown at right. Note that extra glue has been added to form a funnel at the top, allowing saltwater to be added.

Not shown in the photo is a generous additional bead of glue all around the outside of the plates, which is there to ensure that no saltwater will leak out. Allow the glue to cure before going on to the final step.

Last, use a large eyedropper to add saltwater to the cell. Fill it up almost to the top of the copper plate so that it almost spills out. Then seal the funnel with another generous bead of glue. Let the glue cure at least a half hour.

In the photo at left above, you can see the flat panel solar cell in action in the bright sun. It is delivering about 36 microamperes of current. You can also see the extra bead of glue around the edges of the plates, and filling the top of the funnel.

Finally, at right, another shot of the author's shadow. Note that the meter now reads about 4 microamperes, since no sunlight is falling on it.

RADIOS

A crystal radio is the distilled essence of a radio. It has very few parts, it needs no batteries or other power source, and it can be built in a short time with objects you can find around the house or at your local store.

The reason a crystal radio does not need any batteries is because of the amazing capabilities of the human ear. The ear is extremely sensitive to very faint sounds. The crystal radio only uses the energy from radio waves sent by radio transmitters. These radio transmitters send out enormous amounts of energy (tens of thousands of watts). However, because they are usually far away, and we have at most a few hundred feet of wire for an antenna, the amount of energy we receive with the crystal radio is measured in billionths of a watt. The human ear can detect sounds that are less than a millionth of that.

To start this chapter, you'll build a simple working radio using parts that you can purchase at stores like Radio Shack, or through mail order. You will use common household objects when you can, but your emphasis will be to quickly put together a radio that works.

Later you will learn more about radios by looking at even simpler versions that might not work as well as our first radio but will show the important radio concepts more clearly, since they have fewer parts.

Then you will improve your radio, making it louder, allowing it to receive more stations, and improving its looks. Lastly, you will build each part of the radio from scratch, using things you can find around the house. This will take a lot longer than your first radio, but since it can be done by replacing store-bought parts one at a time, you will always have a working radio.

Building a Simple Crystal Radio

SHOPPING LIST

- ◌ 50 feet of enamel-coated magnet wire (Most common gauges—wire diameters—will work, but thicker wire is easier to work with, something like 22 gauge to 18 gauge. You can buy it at Radio Shack [RS# 278-1345], or you can take apart an old transformer or electric motor that is no longer needed and use its wire. You can also use vinyl-coated wire such as RS# 278-1217, which in some ways is easier to use than enamel-coated wire, since it is easier to remove the insulation.)

- ◌ A germanium diode (Most stores that sell electronic parts have these. They are called 1N34A diodes [RS# 276-1123]. These are better for our radio than the more common silicon diodes, which can be used but will not produce the volume that germanium diodes will.)

- ◌ A telephone handset (You will listen to this radio in the same way you listen to the phone. If you have an old telephone sitting around, or can find one at a garage sale, you're set. Or you can buy the handset cord [RS# 279-316] and borrow the handset from your home phone; using it for the radio will not harm it.)

ᴐ Set of alligator jumpers [RS# 278-1156], or you can find them anywhere electronics parts are sold.

ᴐ About 50–100 feet of stranded insulated wire for an antenna (This is actually optional, since you can use a TV antenna or FM radio antenna by connecting your radio to one of the lead-in wires. But it's fun to throw your own wire up over a tree or on top of a house, and it makes the radio a little more portable.)

ᴐ Sturdy plastic bottle (I have used the plastic bottle that hydrogen peroxide comes in, or the bottles that used to contain contact lens cleaner. They are about 3 inches in diameter and 5 to 7 inches long. Shampoo bottles also work, but you will want to get the ones with thick walls, rather than the thin, flimsy ones. This will make it easier to wind wire around them.)

TOOLS

☐ Nail or ice pick
☐ Soldering iron
☐ Tape

Use a sharp object, like a nail or an ice pick, to poke four holes in the side of the plastic bottle. Two holes should be about a half-inch apart, near the top of the bottle, and should be matched at the bottom of the bottle with two more just like them, as shown. These holes will hold the wire in place.

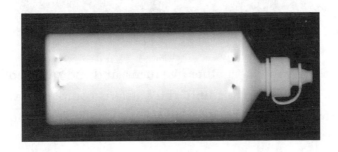

Thread the magnet wire through the two holes at the top of the bottle, and pull about 8 inches of wire through the holes. If the holes are large and the wire is loose, it is OK to loop the wire through the holes again, making a little loop of wire that holds snugly.

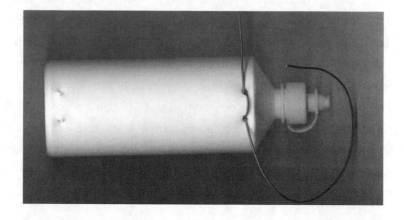

Now take the long end of the wire and start winding it neatly around the bottle. When you have wound five windings on the bottle, stop and make a little loop of wire that stands out from the bottle. It's easy if you wrap the wire around a nail or a pencil.

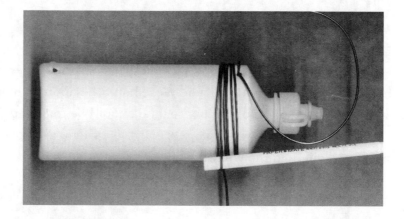

Continue winding another five turns, and end with another little loop. Keep doing this until the bottle is completely wrapped in wire and you have reached the second set of holes at the bottom of the bottle.

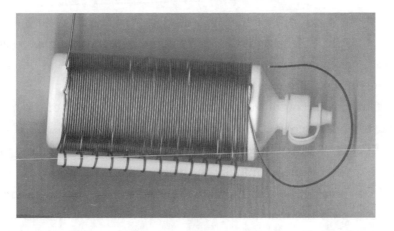

Cut the wire so that at least 8 inches remain, and thread this remaining wire through the two holes just as you did at the top of the bottle. The bottle should now look like this:

Now remove the insulation from the tips of the wire, and from the small loops (called "taps") you made every five turns. If you are using enameled wire, you can use sandpaper to remove the insulation. You can also use a strong paint remover on a small cloth, though this can be both messy and smelly. Don't remove the insulation from the bulk of the coil, only from the wire ends and the small loops. If you are using vinyl-coated wire, the insulation comes off easily with a sharp knife.

Next, attach the germanium diode to the wire at the bottom of the bottle. It is best to solder this connection, although you can also twist the wires together and tape them, or you can use alligator jumpers if you are really in a hurry.

Cut one end off of the handset cord to remove one of the modular telephone connectors. There will be four wires inside. If you are lucky, they will be color-coded; you will use the yellow and black wires. If you are unlucky, the wires will be all one color, or one will be red and the others will be white. To find the right wires, first strip off the insulation from the last half-inch of each wire. Then take a battery, such as a C, D, or AA cell, and touch the wires to the battery terminals, one wire to plus and another to minus, until you hear a clicking sound in the handset earphone. When you hear the click, the two wires touching the battery are the two that go to the earphone. These are the ones you want.

The "wires" in the handset cord are usually fragile copper foil wrapped around some plastic threads. This foil breaks easily, sometimes invisibly, while the plastic threads hold the parts together making it look like there is still a connection. I recommend carefully soldering the handset wires to some sturdier wire, then taping the connection so nothing pulls hard on the copper foil.

Attach one handset wire to the free end of the germanium diode. Solder it if you can.

Attach the other wire to the wire from the top of the bottle. Soldering this connection is a good idea, but it is not necessary.

Clip an alligator jumper to the antenna. Clip the other end to one of the taps on the coil. Then clip another alligator lead to the wire coming from the top of the bottle. This is your ground wire, and it should be connected to a cold water pipe or some other metal object or wire that has a good connection to the earth.

At this point, if all went well, you should be able to hear radio stations in the telephone handset. To select different stations, clip the alligator jumper to different taps on the coil. In some places you will hear two or more stations at once. The longer the antenna,

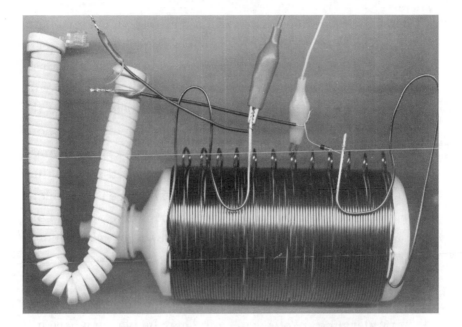

the louder the signal will be. Also, the higher you can get the antenna the better.

Now that you have a working radio, you can make it look more attractive and sturdy by mounting it on a board or in a wooden box. Machine screws can be stuck into holes drilled into the wood to act as places to attach the wires, instead of soldering them. A radio finished this way is shown at right. Note the nice little touch of using brass drawer pulls on the machine screws to hold the wire.

What Do I Do If My Radio Doesn't Work?

Here are some things that might have gone wrong:
- You may have a low-impedance earphone. This means that the number of windings in its coil is too small, and it acts like a short-circuit to the radio. You can replace it with a high-

impedance earphone or you can use an impedance-matching transformer to match the radio to your earphone. An imped-ance-matching transformer has one side with lots of turns on its coil, and another with fewer. You connect the side with lots of turns (an impedance of 1,000 or 2,000 ohms) to the radio in place of the earphone. You then connect the earphone to the side with fewer turns (between 4 ohms and 32 ohms).

- You may not have a long enough antenna. Anything less than 30 meters long is probably inadequate. If space is a problem, try winding the antenna around the room in a big coil; how-ever, having the antenna outdoors, as high up as possible, is best.

- You may have a bad ground connection. A cold water pipe is usually used, since one end of it is usually buried. (Not so for hot water pipes.) Your water pipe may be plastic where it is buried, which would not be a good ground. As an alterna-tive, find some metal that is buried in the ground at one end, and connect the ground of the radio to that.

- You may be far away from a strong radio station. If so, you will need a larger antenna. If you have access to an amplifier, such as a stereo system with a phonograph input, try con-necting the phonograph input of the amplifier instead of the earphones, so you can amplify any signal you get. If you still get no signal, then the problem is *not* your earphone or your antenna and may be your connections or your diode. If you are having trouble finding good diodes, try using any two wires from a three-wire transistor.

- Check your wiring carefully against the schematic diagram to make sure that all your connections are making good contact.

Building a Radio Out of Household Implements

A Piezoelectric Earphone

The most difficult part of building a crystal radio is building an efficient earphone that can convert the tiny electrical signals into sounds your ears can hear. The first radio you built used a telephone handset for an earphone, and that works quite well. But another type of earphone is available that fits into your ear so you don't have to hold it. It is also more sensitive than the telephone handset.

In order to convert very faint electrical signals into sound, you need a very sensitive earphone. The kind of earphones used in transistor radios or CD players will not do; those earphones are meant to be driven by a signal loud enough to drive a speaker, and are not sensitive.

You will learn later in this chapter about impedance. For now, just know that a sensitive earphone has a very high impedance, which is measured in ohms. A speaker has a low impedance, usually about 8 ohms. A sensitive earphone built around an electromagnet might have 2,000 ohms. The telephone handset earphone is built around an electromagnet, although it has only a few hundred ohms of impedance and will not be as loud as a more sensitive device.

The crystal earphone you will work with in this section—more properly called a piezoelectric earphone—has over a million ohms of impedance and is very sensitive. You can purchase a piezoelectric earphone from an electronics parts supplier or from www.scitoys.com for about $5.00.

A crystal earphone is made of a material that changes its shape when connected to a source of electricity. Some crystals, such as quartz and Rochelle's salt, are piezoelectric. Some ceramics, such as those made with barium titanate, are also piezoelectric. Your piezoelectric earphone is made of a disk of brass that is coated with barium titanate ceramic. When electricity is connected to it, the ceramic bends the brass disk, and you can hear the vibrations this causes in the air.

To demonstrate just how sensitive a crystal earphone is, try this experiment: with the earphone in your ear, touch the two wires together. You will hear a sharp click as electrons move from one wire to the other. If the earphone came with a jack on the end instead of two bare wires, you will need a piece of metal, such as a spoon, to connect the two metal parts of the jack.

One detail about such a very sensitive earphone is important in building a crystal radio. A sensitive earphone does not use much current to create sounds. Your radio needs a certain amount of current to flow through the diode in order to work.

When substituting a piezoelectric earphone for an earphone made with a coil of wire, you must provide a way for some current to bypass the earphone. You can do this by putting a resistor or a coil in parallel with the earphone. Parallel means that the resistor or coil is attached to the same two places that the earphone wires are attached. The resistor can be anything in the range of 1,000 ohms to 100,000 ohms and can be a piece of graphite from a pencil, or a few hundred coils of fine wire around a nail.

A Germanium Diode Detector

The second part of your new radio is the detector. A detector picks the audio frequencies out of a radio wave so they can be heard in the earphone. You will learn more about how a detector works later in this chapter.

Your first detector will be store-bought. Later you will replace it with detectors you build out of things you find around the house, such as graphite pencils, baking soda, razor blades, rocks, all kinds of things.

The detector you will use first is a germanium diode. The diode you need is a 1N34A, which has some properties that make it particularly suited to this project, namely that it works at lower voltage levels than most  other common diodes. Since the voltage in your radio comes from weak radio waves, you need all the help you can get.

You are now ready to build the simplest radio.

A Very Simple Radio with Two Parts

First let me warn you that this first radio may not work in your location. You must have a very strong local radio station to overcome the limitations of such a simple radio. If this radio does not work where you are, you can either skip ahead to the next radio or drive closer to a local radio station and try it there. However, because it is so simple, it's worth building just to see what you might be able to pick up.

If your earphone has a jack on the end, cut it off so there are two long wires coming from the earphone. If the wires are twisted around each other, that is OK, as long as they are separate at the ends. Remove the covering, called insulation, from the ends of the wires to expose an inch of bare wire. Often you can do this with your fingernail, but a tool called a wire stripper is made for this purpose and is usually available at the same place you bought the earphone or the diode.

Wrap one bare wire around one of the diode's wires. Use some tape to keep it in place. If you know how to solder, you can solder the wires together, but it really isn't necessary for now.

Tape the other diode wire to a cold water faucet. This makes a good ground connection.

Hold the remaining free bare wire of the earphone in your hand. This makes your body a radio antenna. Put the earphone in your ear. If you are close to a strong AM radio station, you will be able to hear that station faintly in the earphone. You may hear more than one station at once.

If you can't hear anything, try a better antenna. You can tape the wire you were holding to a metal window screen, or a long wire. If one end of the long wire is thrown up on a roof or in a tree, you might get better results. Another good antenna is an outdoor TV antenna. Just touch the free earphone wire to one of the antenna terminals where it comes into the TV. If you have a good antenna you may be able to eliminate the ground connection, using your body as a ground instead by holding the free diode wire in your hand.

A Simple Radio with Two Parts

Your simple radio has two main drawbacks. One is that the signals are very faint and can only be heard if you are close to a radio station's transmitting antenna. The other is that you hear all of the strong stations at once, and it is hard to pick out just one song or voice from the jumble. The first problem refers to the *sensitivity* of the radio; this radio is not very sensitive. The second problem refers to the *selectivity* of the radio; this radio is not very selective. You can solve both problems by using a trick called *resonance*.

Resonance is a way of taking a little bit of energy and using it over and over again, at just the right time, to accomplish a big task. Resonance is the same strategy you use when you push someone on a swing. It would take a lot of work to lift someone several feet in the air, but you can do this easily on a swing by giving a little push over and over again at just the right time. But timing is important: if you push at the wrong time, the swing can actually lose energy instead of getting higher.

When an opera singer uses her voice to shatter a wine glass, she is using resonance. Her voice gives the glass a little push at just the right time, over and over again, until the glass is moving so far that it shatters. In a similar way, you can splash all the water out of a bathtub by moving your hand back and forth in the water at just the right speed. Each time your hand moves, the water climbs a little higher until it is over the top of the tub.

Radio waves can act like the sound waves of the singer's voice, or like the waves in the bathtub. Radio waves can cause electrons to move back and forth in a wire, just like the water in the tub. If the radio waves move back and forth at the right frequency, the electrons in the wire will crowd toward one end of the wire when the radio waves move them back to the other side. Just like the water in the tub, the electrons will crowd together higher and higher at the ends of the wire. These electrons can do work, like moving the brass disk in the earphone to create sound.

You can use resonance to build a radio that can pick up only one station at a time, and make a louder sound in the earphone. However, this radio will have a drawback—it will be over 1,000 feet long! But you will solve this problem in the next radio you build.

Suppose you pick a local radio station you want to hear. For this example, assume you want to listen to 740 kilohertz on the AM dial. You now need to figure out how long the wire must be to resonate at this frequency. Radio waves travel at the speed of light. This radio wave is going back and forth 740,000 times per second. This means the wave needs to go about a quarter of a mile in one direction, then turn around and go back again, over and over. The formula for determining how long the wire should be is:

$$\frac{936 \text{ feet}}{\text{frequency in megahertz}}$$

or, for our example:

$$\frac{936 \text{ feet}}{0.740 \text{ MH}}$$

or about 1,264 feet.

To make this radio, take half of the wire (632 feet) and attach it to one end of the diode. Attach the other half of the wire to the other end of the diode. Attach one earphone wire to one side of the diode, and the other earphone wire to the other end. Put the long wire up in the air by attaching each end to a tree. The trees should be about 1,264 feet apart. Then put the earphone into your ear and listen to the radio.

Now I can think of a few problems with this radio. It is not very portable. Also, in order to change the station, we need to make the wire longer or shorter. One solution to the portability problem is to coil the wire up by winding it on a box or a cylinder. Then you can solve the tuning problem by attaching the diode and earphone to the coil at different places, which is easy to do now that the whole wire is in one small place.

A Simple Radio with Three Parts

There are several ways to connect a coil of wire to a diode and earphone to make a radio. The photos below show two possibilities.

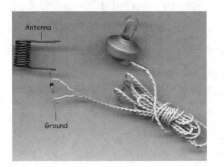
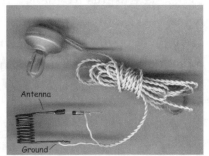

These photos do not show the antenna and ground connections, but instead indicate where they would be attached.

The coils in the photos are also dramatically simplified. A real coil for the AM radio frequencies would be somewhat larger, as you saw when you built your first radio using the plastic bottle.

Often photographs show so much detail that the important parts are easily missed. By using a simplified diagram, you can accentuate the important parts of the circuit and leave out unimportant or distracting details. A simplified diagram of a circuit is called a schematic. A schematic for a simple crystal radio might look like this:

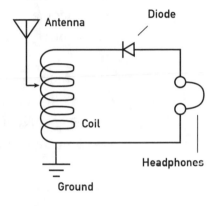

The symbol for a coil looks like a spring. The symbol for an antenna looks like a coat hanger. The symbol for headphones looks like old-fashioned earmuffs (which are great for crystal radios, since they block out ambient noise in the room). The symbol for the ground looks like what a cartoonist would draw under a cartoon character to represent the earth.

Note that the antenna is attached to the coil in the middle by a small arrow. This indicates that it is attached to a tap in the coil. An arrow is used to indicate a connection that can move, like our clip lead.

The symbol for the diode looks nothing like the little glass tube with wires coming out. Instead of representing what the diode looks like, it represents what the diode does. A diode is a one-way valve for electricity. The electric current flows through the diode in one direction but is blocked if it tries to flow in the other direction. You will find out why this is important later, when you learn why the radio works. But for now, concentrate on building a radio that will let you hear one station at a time, with reasonable loudness.

Power from Radio Waves—
Measuring the Voltage and Current

It is useful to be able to measure the effects of changes you've already made to the radio. You can just use your ears and try to remember how loud it used to be, but it is easier (and more accurate) to read a meter and remember a number. With a meter connected to the radio you can adjust the tuning to obtain the highest meter reading, or make other adjustments as you add new components or replace purchased components with ones you make yourself.

The meter must be sensitive to very small changes in the amount of electricity flowing in the radio. You will mainly be measuring current, but you will add a voltmeter as well so you can calculate the total amount of energy the radio is receiving.

Current is the flow of electricity through the circuit, and it is measured in amperes, or amps. Voltage is the pressure that pushes the current through the wires. If electricity were water, current would be the amount of water flowing (gallons per minute), and voltage would be the water pressure in pounds per square inch.

Since the amount of current is very small, we will use a meter that measures current in microamperes, or at most small fractions of a milliampere. Some examples of microammeters and milliammeters are shown below:

To measure the current in the radio, you must have the current flowing through the meter. To do this, connect the micro-ammeter between the earphone and the ground connection, so that any electricity that is going to flow through the earphones to make noise is also going to have to flow through the meter. The meter can be connected in two ways, forward and backward. If the meter is connected backward, the needle will start reading below zero. If this happens, just reverse the connections so the needle reads above zero.

To measure the voltage, connect the meter to both of the earphone wires. The schematic now looks like this:

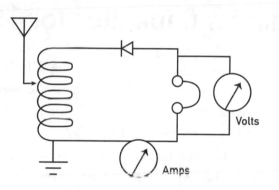

If you have a good antenna or a strong radio station nearby, the ammeter might read more than 50 microamps. If you have a short antenna, you might get only 5 microamps and still be able to hear the station clearly in the headphones. I put up a 200-foot antenna between two trees over my house and tuned to a 50,000-watt news station about 30 miles away, and now I get 175 micro-amps of current through my meter. I put the earphone to the mouth of a cone (like a megaphone) and I can clearly hear the radio from across the room when the house is quiet. It doesn't sound as nice and clear as it does with the earphone right up to my ear, but I can follow the radio conversation easily.

The voltmeter in the same radio reads 125 millivolts. Since watts, the measure of how much power we have, is the voltage multiplied by the amperes, you have 0.000175 x 0.125, or 0.0000218 watts, or about 22 microwatts. The station is putting out 50 kilo-watts, and you are receiving one ten-billionth of that power, yet you can hear it across the room.

Try different lengths of antenna and watch the current go up as the longer antennas catch more of the power from the radio station. Try more than one antenna. Try connecting the ground wire to different things that are connected to the ground, such as pipes, metal fences, etc. With each test, tune the radio again, because your changes may affect the tuning.

Adding a Capacitor (or Three)

As you tried different antenna lengths, you may have noticed that you had to move the tap on the coil in order to receive the station at its loudest. To understand why this happens, and how you can use it to improve your radio, you must first understand capacitance and how it affects the tuning coil.

A *capacitor* is simply two pieces of metal with an insulator between them. If a capacitor is connected to a battery, the battery will push electrons onto one piece of metal, called a *plate*, and draw electrons from the other piece of metal. If you remove the battery the electrons can't go anywhere, so one plate of the capacitor will have more electrons than the other plate.

If you connect the two plates together with a wire, the electrons will rush from the plate that had too many to the plate that had fewer electrons, because electrons have the same charge and thus repel each other like the north poles of two magnets. As the electrons rush from one plate to the other, you can make them do work, such as light a lightbulb. In this way, the capacitor seems to store the electricity from the battery for use at another time when the battery isn't there.

Now suppose you connect a coil and a capacitor together like this:

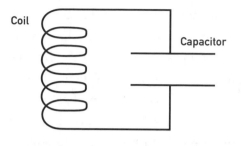

Suppose, too, that the capacitor has been charged by a battery, so the top plate has more electrons than the bottom plate. When you connect the coil, the excess electrons in the top plate immediately start traveling through the coil to get to the plate that has a shortage of electrons.

As the electrons travel through the coil they create a magnetic field (remember "coil" is just another word for "electromagnet"). The magnetic field grows until the plates on the capacitor have equalized. At this point you might think the current would stop flowing in the coil. But the magnetic field that built up when the current flowed through the coil now starts to collapse.

Just as moving a magnet past a coil will generate a current, a collapsing magnetic field around a coil creates a current, too. The current moves in the same direction as it did when the magnetic field was created, so the coil ends up pushing electrons onto the bottom plate of the capacitor and stealing them from the top plate. By the time the magnetic field around the coil has completely collapsed, the bottom plate of the capacitor has a surplus of electrons, and the top plate has a deficit.

You can probably guess what happens next. The electrons start flowing back into the coil, this time from the bottom plate to the top. The coil starts building up a magnetic field again, but since the current is now going the other way, what was once the north pole of the magnetic field is now the south pole, and vice versa. The field grows until the capacitor has equalized, then it collapses and pumps electrons into the top plate of the capacitor. You are now back where you started, and the whole process starts over again.

The coil and the capacitor are resonating, just like the child on a swing or the water in a bathtub. In fact, this circuit is called a "tank circuit," like a tank filled of water that sloshes back and forth.

You can control the frequency of the oscillations in two ways. You can make the coil larger or smaller, or you can make the capacitor larger or smaller. The coil you built for your radio has taps, which have the effect of making the coil shorter or longer, depending on which tap you connect to the antenna.

Your radio has a coil, but it doesn't have a capacitor. Or does it? Actually, the antenna itself is acting like a capacitor. The capacitance of the antenna is reacting with the inductance of the coil to resonate at the frequency of the radio station.

When you change the length of the antenna, it is like changing the size of the capacitor. This is why changing the length of the antenna changed the tuning of the radio, forcing you to move to a different tap on the coil in order to listen to the same station.

There is another way to change the capacitance of a capacitor: you can change the distance between the two plates. If the plates are closer together, the excess electrons on one plate are attracted to the other plate because when the negatively charged electrons were removed from that plate, it was left with a positive charge.

Because the electrons are attracted to the positive charge, you can pile more of them together, storing more energy. In a similar fashion, when you make a capacitor with the plates farther apart, the positive charge is farther away and can't help to pull as many electrons onto the negative plate. Thus the amount of energy you can store is less, and the capacitor has less capacity.

You can combine capacitors to raise or lower the capacitance, now that you know how capacitors work. If you put two capacitors together in parallel you can increase the capacitance, because the top plates are connected together and the bottom plates are connected together; it is as if you had one capacitor with large plates.

If you connect the capacitors in series, the plates of the capacitors will be farther apart. This is shown in the top diagram on at right. The bottom plate of one capacitor is connected to the top plate of the other. Electrically, this is the same as making the two plates into one plate in the middle of a capacitor that has twice the distance between the outer plates. The phantom inner plate has no effect and is drawn as a dotted line in the bottom diagram.

You now know enough about capacitors to use them in your radio. You can use a small capacitor between the antenna and the coil to lower the capacitance of the antenna. This will allow the coil to tune to stations that are higher in frequency. The capacitor is in series with the capacitance of the antenna, so the total capacitance is lower. The circuit now looks like this:

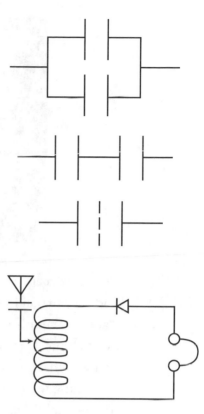

Building Your Own Capacitors

Capacitors are easy to build in the kitchen out of aluminum foil. In fact, your first capacitor will simply be two sheets of foil tucked into a paperback book with one page separating them, as if they were two bookmarks.

SHOPPING LIST
- Aluminum foil
- Waxed paper or plastic wrap

TOOLS
- Scissors

This quick capacitor has advantages and disadvantages. It is quick and easy to build, and it can be easily adjusted to reduce the capacitance by simply sliding one of the foil strips out of the book a little at a time. On the other hand, it is bulky and comes apart easily, and will change its capacitance when you press down on the book, squeezing the pages closer together. Lastly, it can change capacitance slightly on humid days as the pages of the book absorb moisture.

With only a little more effort, you can make a durable, stable capacitor using foil and a little waxed paper or plastic wrap. Start by laying down a sheet of waxed paper. Lay a sheet of foil on top of that. Leave the foil hanging over the top of the waxed paper, so that you will have something to which you can attach a wire. Lay another piece of waxed paper over the first piece and the foil, then lay another piece of foil on the top, overlapping it at

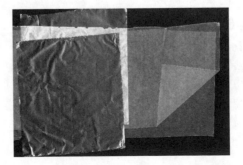

the bottom for your other wire. Make sure that the foil sheets are always separated by the waxed paper, so they do not make an electrical connection.

Now roll the whole thing up like a jelly roll, as shown at right.

Trim the paper with some scissors. You can even fold it up the other way to make it smaller.

This capacitor is not adjustable like the first one. You can make several of them, each a different size, and connect the one you want. You can even combine them in parallel or in series to change their capacitance.

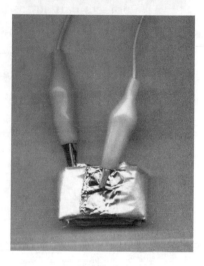

You can use the small fixed capacitor to tune the antenna, and another variable capacitor (like the book capacitor) to tune the coil. Put the variable capacitor in parallel with the coil to make a tank circuit. The small fixed capacitor lowers the antenna's capacitance, making the circuit tune to a higher frequency. But the variable capacitor adds more capacitance to the circuit, making it tune to a lower frequency. Now you can tune the radio with the taps on the coil, and by sliding the foil in and out of the book.

The circuit now looks like the diagram at right.

Notice how the variable capacitor has an arrow through it to indicate that it can change its capacitance.

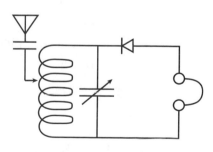

Building Your Own Diodes

During World War I, soldiers in the field made their own radios to listen to programs for entertainment and news. They had access to wire from broken-down vehicles and telephone receivers, but they did not have modern solid-state diodes in little glass tubes.

However, it is surprising to find out just how many ordinary objects can act as a diode, letting current flow one way better than another. The soldiers found that an old rusty razor blade and a pencil lead worked just fine. By lightly touching the pencil lead to spots of blue on the blade, or to spots of rust, they formed what is called a *point contact diode*.

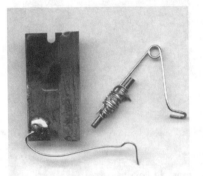

You can replace your store-bought diode with a homemade point contact diode and compare the results. The parts can be attached to the circuit with clip leads, or they can be soldered as in the photo at left. The pencil lead is attached to a safety pin by wrapping it with bare copper wire and soldering it.

SHOPPING LIST

- ꝰ Wire
- ꝰ Pencil lead
- ꝰ Clip leads
- ꝰ Brass drawer pulls
- ꝰ Razor blade
- ꝰ Safety pin

TOOLS

- ◻ Soldering iron

The safety pin acts as a spring to lightly press the pencil lead onto the razor. If the pressure is too hard or not hard enough the diode will not work, so experiment. The exact spot on the razor is also critical, since some spots will have too much or too little oxide on them to make the diode. Move the pencil lead around on the razor until the sound is loudest, or the meter (if you have attached one) reads highest.

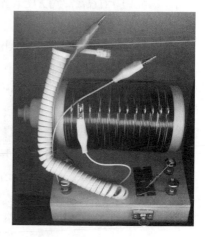

In the photo at right, you can see how handy the brass drawer pulls are when you want to attach new types of diodes.

If you don't have a rusty razor blade lying around, you can try other bits of rusty metal. The blade shown above was clean and new, so I put a little salt and water on it and held it in the flame of a gas stove until parts of it were blue and purple.

You might have other things around the house that can act as diodes. In my rock collection I found some iron pyrite (fool's gold) and some carborundum (silicon carbide, the stone shown at right). Carborundum works well using a strong pressure, so I simply wrapped some bare copper wire around it, soldered the wire, and let the jaws of a clip lead supply the pressure. It worked quite well. The pyrite needed a gentle touch, so I used the point of a safety pin to gently probe until I found a spot on the pyrite that gave good volume in the radio.

Going Further

Trading Loudness for More Stations

In your radio, the diode and earphones are connected directly to the antenna and ground. This connection gets the loudest signal. However, it also loads the tuning coil, making it less selective. This means that local strong stations drown out many lower power or distant stations.

You can make the radio more selective by decoupling the tuning coil from the antenna and ground by adding a small coil. The new coil is attached to the antenna and the ground, and then it is placed inside the main tuning coil.

SHOPPING LIST
- Wire
- Plastic container (such as a 35-millimeter film container about 1 inch in diameter)

TOOLS
- Scissors

Wind about 5 or 10 turns of wire around a small coil form such as the plastic container used to package 35-millimeter film (about 1 inch in diameter). Cut a large hole in the bottom of the plastic bottle on which you wound the large tuning coil. Attach the antenna and ground to the small coil, and place it into the large tuning coil using the new hole you just made. By moving the small coil in or out of the large coil, you can vary the coupling between the coils and thus vary the selectivity and sensitivity of the radio. If you want loud strong local stations, place the coil all the way in. If you want to hear the fainter distant stations, pull it out a bit.

Building Your Own Earphones

SHOPPING LIST
- Aluminum can
- Nail
- Small magnet
- Fine wire

You can build your own earphones using an aluminum can, a nail, a small magnet, and some fine wire. Wind a few hundred turns of wire around the nail. Let the magnet stick to the head of the nail (a neodymium-iron-boron supermagnet from Radio Shack works well here, since it is strong and very small). Attach the coil to the radio in place of the earphones. Hold the open end of the aluminum can to your ear and hold the nail very close to the bottom of the tin can. The bottom of the can will be attracted to the magnet, but the coil will make it vibrate with the sound from the radio.

A coil from an old relay or solenoid will also often work, and it will save you the effort of winding the wire on the nail.

A Seashell Loudspeaker

I once bought a large conch shell from an aquarium store for a few dollars. Using a concrete drill, I made a ¼-inch hole in the shell at the small end (where the shell was formed when the conch was very small). I then glued a piezoelectric earphone to the hole. This made a nice trumpet-like megaphone and made the sound of the radio clearly audible across a quiet room.

Using an LED for a Diode

Because I have a long (150-foot) antenna, a good ground, and a strong station (50,000 watts) less than 20 miles away, my radio receives enough power to light a low-current LED. The LED is a "high brightness" type, which also means that it will light dimly with a very small amount of current. I connected it instead of a diode in the radio, and it glows as the radio operates, getting brighter as the sound gets louder.

If you don't have a strong station nearby, you can add a battery in series with the LED—a small 1.5-volt battery works fine. The LED will light up, and the radio will play much louder than without the battery. If the LED doesn't light up, try connecting the battery the other way. This arrangement is the best detector I have used so far and is louder than the 1N34A germanium diode.

Building a Radio in 10 Minutes

For our 10-minute radio, we will need a few parts. (A bundle of all the necessary parts is available at www.scitoys.com.)

SHOPPING LIST

- ➲ Ferrite loop antenna coil (In your other crystal radios you wound the coil by hand. In this project you use a much smaller coil with a ferrite rod inside, from www.scitoys.com. The ferrite rod allows the coil to be smaller, and it can be moved in and out of the coil for coarse tuning.)
- ➲ Variable capacitor, 30–160 microfarads (Available from www.scitoys.com, or you can find them in old broken or discarded radios.)
- ➲ Germanium diode (1N34A)
- ➲ Piezoelectric earphone

- A short wire or alligator jumper (Use an alligator jumper here for convenience, RS# 278-1156, or you can find them anywhere electronics parts are sold. You can use any piece of insulated wire instead.)
- 50–100 feet of stranded insulated wire for an antenna (This is actually optional, since you can use a TV antenna or FM radio antenna by connecting your radio to one of the lead-in wires. But it's fun to throw your own wire up over a tree or on top of a house, and it makes the radio a little more portable.)
- Block of wood or something similar for a base
- Solder

TOOLS
- Soldering gun
- Tape

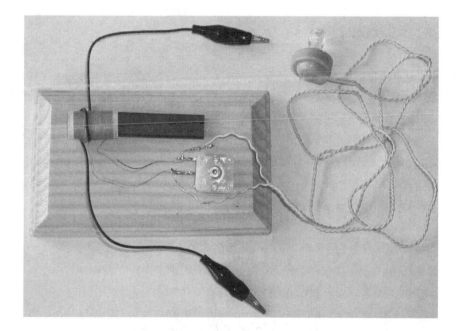

You can see from the photo above how simple this radio is and why it can be put together in a very short time.

Solder the wires from the ferrite loop to the two rightmost leads of the variable capacitor. It doesn't matter which wire goes to which lead.

Solder the germanium diode to one of the leads of the variable capacitor. Again, it doesn't matter which lead.

Solder one of the piezoelectric earphone wires to the free end of the germanium diode. Solder the other to the lead of the variable capacitor that does not have the diode attached to it.

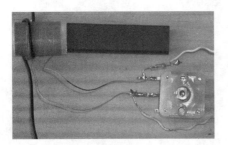

Last, loop the alligator jumper around the coil once (if you have a long antenna and a good ground) or a few times (for shorter antennas). Glue the coil and the wire down later, after you have tested the radio. In the meantime, use some sticky tape to hold it onto the base.

And that's it—you're done!

How Does It Work?

To use the radio, connect one end of the alligator jumper to your antenna. Connect the other end to a good ground, such as a cold water pipe.

Start tuning with the variable capacitor set in the middle of its range, neither all the way clockwise nor all the way counterclockwise. With the earphone in your ear, slowly move the ferrite rod into the coil, listening for radio stations.

With a long antenna, and a single loop of the alligator jumper, you can tune several radio stations. In some areas, one or two stations will be so close or so powerful that they overwhelm all the others, so you will only hear those one or two stations.

If you have a shorter antenna, the stations will sound very faint if you only use one loop of the alligator jumper. Looping the jumper around the coil two or three times will increase the volume. But the volume increase comes at a price; you will hear fewer stations.

The number of loops in the alligator jumper trade off volume for selectivity. The best way to increase the volume is to use as long an antenna as you can and a good ground connection.

How Does the Ferrite Change the Frequency?

The ferrite rod increases the inductance of the coil. In your other (hand-wound) coils, you increased the inductance by winding some more loops, or by using a "tapped" coil and selecting a tap that was farther down the coil.

As the ferrite rod is inserted into the coil, more of the coil is affected by the ferrite, and so the inductance increases. Increasing the inductance moves the frequency lower. This allows you to hear stations "lower on the radio dial."

Ferrite is used because it is magnetic, like iron or steel, but it is not a conductor of electricity. If it were conductive the coil would induce "eddy currents" in it, and some of the energy would be lost heating up the core. Because ferrite is not a conductor, we can use its magnetic properties to change the inductance of the coil without losing volume.

If you have a long antenna, a good ground, and you are not too close to a strong station, the variable capacitor will help in fine tuning the stations.

Building a Very Simple AM Voice Transmitter

If a crystal radio is the distilled essence of a radio, this simple transmitter is the distilled essence of radio transmitters.

The transmitter goes together in about 10 minutes, and is small enough to fit in the palm of your hand. Depending on the antenna, the transmitter can send voice and music across the room or across the street.

I put together my first version with simple clip leads—no soldering, no printed circuit board, not even a battery clip. This version is much sturdier and convenient.

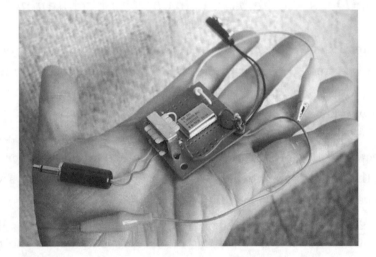

SHOPPING LIST

◐ 1-megahertz crystal oscillator (This is a crystal clock oscillator such as those used in computers. There are many suppliers, such as Jameco [#27861], JDR MicroDevices [#OSC1.0], or www.scitoys.com.)

◐ 1000-ohm to 8-ohm audio audio transformer (RS #273-1380)

◐ Generic printed circuit board (I used Radio Shack's #276-159A, but any general-purpose printed circuit board will do.)

◐ Phone plug (This should match the jack in your sound source. I use a ⅛-inch plug [RS #274-286A] to match standard earphone jacks of transistor radios and Radio Shack's Archer mini-amplifier speaker.)

◐ 9-volt battery clip (RS# 270-324)

◐ 9-volt battery

◐ Set of alligator jumpers (RS #278-1156)

◐ Insulated wire, for an antenna (You can use the same antenna you used for the crystal radio.)

Building the Transmitter

The oscillator is the heart of the transmitter. It has four leads, but you only use three of them. When the power is connected to two of the leads, the voltage on the third lead starts jumping between 0 volts and 5 volts, one million times each second.

The oscillator is built into a metal can. The corners of the can are rounded, except for the lower left corner, which is sharp. This indicates where the unused lead is. The lead is there to help hold the can down firmly on the printed circuit board, but it is not connected to anything inside the can.

The other main part is the audio transformer. In this circuit it is used as a modulator. The modulator changes the strength of the radio waves to match the loudness of the music or voice we want to transmit.

A pictorial diagram of the transmitter looks like this:

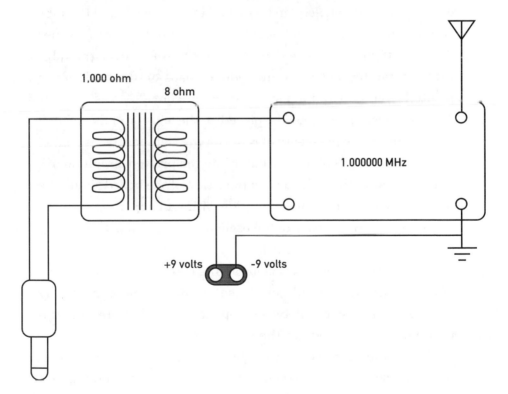

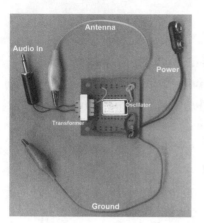

A photograph of the completed transmitter is shown at left:

The transformer has two leads on one side and three leads on the other side. The two leads are the low-impedance side of the transformer, the 8-ohm side. The three leads are the high-impedance side, the 1000-ohm side. The middle of the three leads is called the center tap, and you won't be using it in this circuit.

Putting It Together

The transformer has two metal tabs on the bottom. These can be bent out flat so the transformer can be glued to the printed circuit board, or two holes can be drilled in the board, and the tabs can fit into the holes and be folded over to hold the transformer in place. If you choose to drill the holes and fold over the tabs, the tabs can be soldered to the copper pads on the back of the printed circuit board for a more secure anchor.

The transformer should be placed on the left side of the printed circuit board, leaving plenty of room on the right for the oscillator.

Insert the leads of the oscillator into the printed circuit board, placing it far to the right. The copper side of the board should be down, with the oscillator on the side without copper.

Gently bend the leads of the oscillator over, so it is held firmly onto the printed circuit board.

Solder the pins of the oscillator to the copper foil of the printed circuit board. Be careful not to use too much solder, or it may form bridges of solder between copper traces that are not supposed to be connected together.

Insert the stripped end of the red wire into a convenient unused hole in the printed circuit board (such as the bottom left hole). Insert the red wire from the battery clip into a nearby hole that is connected by copper foil to the first hole, so the two red wires are electrically connected. Solder the two wires to the copper foil.

Insert the white transformer wire into a hole whose copper foil is connected to the upper left pin of the oscillator. Solder this wire to its copper foil.

Cut one of the clip leads in half, so you have two pieces of wire each with an alligator clip attached. I used two different colors for clarity (yellow and green). Strip the insulation from the last half-inch of each piece.

Insert the black wire of the battery clip into a hole whose copper foil connects to the lower right pin of the oscillator. Insert the stripped end of one of the alligator clip leads into a hole that is also connected to the lower right pin of the oscillator. Solder the two wires to the copper foil. The alligator clip will be the ground connection, just like in the crystal radio.

Insert the stripped end of the other alligator clip into a hole that is connected to the top right pin of the oscillator. Solder the wire to the copper foil. This will be the antenna connector.

Open the phone plug and insert the blue and green wires of the transformer into the plastic handle. The metal part of the plug has two pieces, each with a small hole. Put one of the transformer wires into one hole and solder it, then put the other wire into the other hole and solder it. When the metal has cooled, screw the plastic handle back onto the metal phone plug.

Using the Transmitter

You are now ready to test the transmitter.

Plug the phone plug into the earphone jack of a convenient sound source, such as a transistor radio, tape player, or CD player.

Plug the battery into the battery clip.

Hold the transmitter near an AM radio, and tune the radio to

1000 so you can hear your sound source in the AM radio. Adjust the volume controls on the sound source and on the AM radio to get the best sound.

Without any connection to an antenna or a good ground connection, the transmitter will only transmit to a receiver a few inches away. To get better range, clip the ground wire to a good ground, such as a cold water pipe, and the antenna to a long wire like the one you used for the crystal radio. (Many countries limit the length of the antenna you are allowed to use without a license, so check with your local laws before using a wire more than a yard or two long.)

For a science fair project, the transmitter and receiver can be placed within a few feet of one another, and a short wire antenna should be just fine.

☖ WHY DOES IT DO THAT? ☖

The oscillator is connected to one end of a long wire antenna. It alternately applies 9 volts of electricity to the end of the wire and then 0 volts, over and over again, a million times each second.

The electric charge travels up and down the wire antenna, causing radio waves to be emitted from the wire. These radio waves are picked up by the AM radio, amplified, and used to make the speaker cone move back and forth, creating sound.

The sound source (your CD player or tape recorder) is normally connected to drive a speaker or earphone. It drives the speaker by emitting electricity that goes up and down in power to match the up-and-down pressure of the sound waves that were recorded. This moves the speaker in and out, recreating the sound waves by pushing the air in and out of your ears.

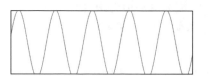

In your transmitter, the sound source is connected to the transformer instead of to a speaker. The transformer is connected to the power supply of the oscillator. The sound source causes the transformer to add and subtract power from the oscillator, just as it would have pushed and pulled on the speaker.

As the power to the oscillator goes up and down, the power of the electricity in the antenna goes up and down also. The voltage is no longer simply 9 volts. It is now varying between 0 volts and 10 volts, because the power from the transformer adds and subtracts from the power of the battery.

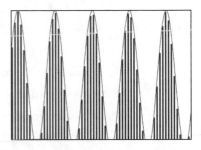

The varying power in the antenna causes radio waves to be emitted. The radio waves follow the same curves as the waves in the antenna. However, because the transmitter and the receiver are not connected, the receiver does not know what the transmitter is using for the value of zero. All that the receiver sees is a radio wave whose amplitude is varying. In the receiver, zero is the average power of the wave. This makes the wave look like the middle drawing at right:

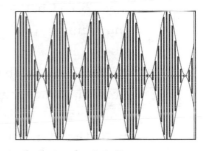

If you sent this wave to the earphone, you would hear nothing because the average power is zero. This is why your crystal radio has a diode.

The diode does a neat little trick. A diode only lets electricity flow in one direction. This means that the part of the graph where the power is rising up from zero can get through the diode, but the part where the power is going down from zero is blocked.

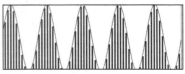

All those little peaks of power happening a million times per second are too fast for human ears and too fast for the earphone to reproduce. But since they are all pushing on the earphone diaphragm, all those little pushes add up, and the earphone moves. Since some of the little pushes are stronger than others (taller bars in the illustration) they move the earphone more than the weaker ones. You hear this variation as sound.

The sound is a faithful reproduction of the original sound wave at the transmitter.

Building a Three-Penny Radio

A crystal radio is nice because it needs no power and the materials can all be homemade, or at least found around the house. But the crystal radio needs a big antenna and a good ground, and it is not very portable.

To get away with using a much smaller, portable antenna, you will need to amplify the tiny signal the radio receives. This requires a portable power supply, such as a battery.

Your next project is a portable radio. It can be powered by a tiny 1.5-volt battery, or from a battery made from copper wire and aluminum foil sitting in a glass of lemonade, a soft drink, or a beer, or by a few small commercial solar cells.

The heart of the radio is a special 10-transistor integrated circuit in a tiny three-legged bit of plastic. This circuit comes ready-made with several amplifiers, the detector, and an Automatic Gain Control circuit that boosts the level of faint stations to match the strong ones, so no volume control is needed. The final radio has excellent performance, pulling in weak stations and preventing nearby strong stations from overwhelming the weak ones next to them on the dial.

It is called a Three-Penny Radio because it uses three shiny pennies as anchors for the various parts the radio needs. This makes the construction very easy.

If you have never soldered anything before, this is a great project to start with. It is very forgiving of the type of soldering usually done by beginners, and all the parts are widely separated, making the job much easier than with other circuits. Soldering irons and solder are inexpensive tools you can find at a local electronics store.

SHOPPING LIST

- Tuning coil (You can wind one by hand, but in this project we use a much smaller coil with a ferrite rod inside.)
- MK484-1 AM radio integrated circuit (This is the heart of the radio.)
- Piezoelectric earphone
- Tuning capacitor (A variable capacitor, from 0 to 160 microfarads.)
- 100,000-ohm resistor (This resistor has four colored bands on it: brown, black, yellow, and gold.)
- 1,000-ohm resistor (This resistor also has four colored bands on it: brown, black, red, and gold.)
- 0.01-microfarad capacitor (This capacitor will be marked something like ".01M" or "103".)
- 2 0.1-microfarad capacitors (These capacitors will be marked something like ".1M" or "104".)
- 1.5-volt battery (optional)
- 1.5-volt battery holder
- 3 shiny pennies (You can polish old pennies or use brand-new ones.)
- Old board
- Plastic lid from a jar or can
- Pushpins

TOOLS

- Small, sharp knife
- Silicone rubber glue

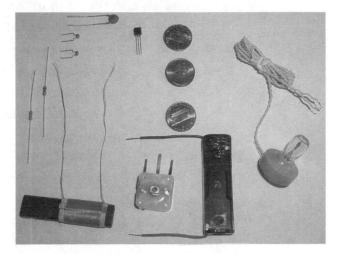

Start by placing the three shiny pennies on an old board where you will work. An old board will not be missed if a hot soldering iron burns a black spot in it. Don't work on a nice tabletop.

The pennies should be clean and bright. This will help the solder stick to them and flow onto their surface. Solder will not stick to a dirty penny. I used clean relatively new pennies that I didn't have to clean or polish. Old pennies can be cleaned with brass polish or by simply leaving them in a mixture of vinegar and salt for a half hour or so.

You are going to build the radio upside down, so that all of your soldering will be neatly hidden from view when you turn the radio over when you are finished. Select which side of the penny you want to be visible, and place that side facedown. I chose "heads" to be visible, so the "tails" side is facing up in the next photo.

First, bend the wires of the integrated circuit so the outside wires stick out like the arms of a scarecrow. This makes the soldering much easier, since the wires are not close together.

The integrated circuit has a flat side and a rounded side. The flat side will face up when you are done, so place it facedown while you build your radio upside down. The orientation of this part is important. From left to right, the three legs are the "output," the "input," and the "ground." (When faceup, the "ground" will be on the left and the "output" will be on the right.) If the

integrated circuit is not flat-side-down at this point, then you won't be connected to the right parts when you are done, and the radio won't work.

It usually takes a while to heat up a penny enough to melt solder onto it. Hold the soldering iron firmly on the spot on the penny where you want the solder to be, and feed the wire solder onto the hot penny as it melts. It doesn't take much solder. It is often a good idea to make a small blob of solder on the penny first, and then place the wire of the integrated circuit onto the blob of solder, and reheat both until the solder wets the wire.

Look at the photo on page 124. The two top pennies have three blobs of solder on them, and the bottom penny has two blobs of solder.

Solder all three of the wires to the three pennies, as shown at top right.

The next step is to solder the variable capacitor to the bottom penny. Remember to place the variable capacitor upside down. The variable capacitor has three legs, but you will only be using two of them. This variable capacitor is actually two capacitors in one, and they share the middle leg.

You will only be using one of the capacitors. The two capacitors have different values, and we are using the 160-picofarad side (the left in the photo) and leaving the 60-picofarad side unconnected. As the photo in the middle shows, the third leg has been cut off to remind you which side to use.

The next part you will add to the circuit is the small fixed value capacitor, the one marked ".01M" or "103." Both of these markings mean the same thing—the capacitor has the value 0.01 microfarads (you could also say 10 nanofarads, but the tendency in the industry is to use microfarads).

Solder the small capacitor to the middle leg of the variable capacitor, and to the penny. It is probably easiest to solder it to the penny first and then bend it so the other leg touches the middle leg of the variable capacitor, and then solder them where they touch. Always make sure metal parts to be soldered are touching before you solder them; this makes a stronger joint.

Next, solder on the 100,000-ohm resistor. In the top photo, you can see the color-coded bands on it. If you could see the colors, they are brown, black, yellow, and gold.

This resistor must be soldered to the middle leg of the variable capacitor at one end, and to the top left penny at the other end. It must not touch any other metal part along the way.

In the top photo, the project has been turned over for a moment to show that the resistor is not touching anything except where it is soldered.

Now solder the 0.01-microfarad capacitor to the two top pennies. This capacitor will be marked "104," or sometimes "0.1M." If the leads are short, the capacitor can be stretched across the integrated circuit as shown in the two middle photos.

If the leads are long, the capacitor can be placed above the integrated circuit. Make sure the wires from the capacitor do not touch the middle wire of the integrated circuit.

Next, connect the wires from the piezoelectric earphone to the 1,000-ohm resistor and to the other 0.1-microfarad capacitor. The color codes on the 1,000-ohm resistor should be brown, black, red, and gold.

Now solder the resistor to the top left penny, as shown in at top left on the next page.

In the next photo, the red (positive) wire from the battery holder is wrapped around

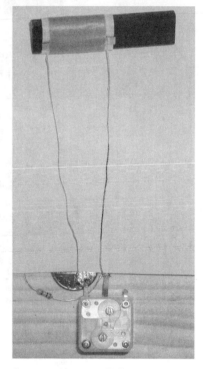

the resistor wire. The black wire goes to the top right penny. Solder all of the connections.

If you are going to use lemonade for the battery, just solder a longish wire to each of these spots instead of the battery holder. I like to use red wire for the positive side and black wire for the negative side, just like they do for the battery holder. This helps me remember which wire goes where later.

The next step is to solder the wires from the coil to the legs of the variable capacitor. In the photo at right, I have placed a piece of white paper over the project, to make it easier to see the fine wires in the photo. (You won't need the paper when you build the radio.)

The ferrite rod in the coil is not glued in place and can slide easily into and out of the coil. This is important, because we will be sliding the ferrite rod in and out of the coil later to adjust the tuning.

The photo above shows the project so far, without the paper in the way.

At this point, the radio is actually complete. You could proba-

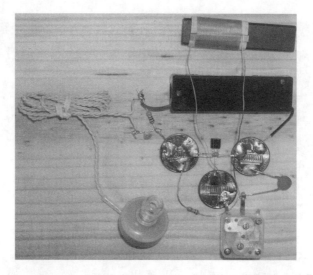

bly hear sounds from the earphone if you put the battery into
the holder. You will learn how to tune the radio in a moment.

Now you finally get to turn the radio over, so it is right-side-
up. The photo shows three pushpins placed around the variable
capacitor. You'll read why a bit later. The capacitor is now glued
down to the board. Notice that the photo shows a nice, clean
board, since all soldering has been completed.

The radio can be used as it is, though you will add a finishing
touch in a moment. It can be tuned in two ways. First, you can
slide the ferrite rod very slowly into and out of the coil. This is a

coarse adjustment, and getting exactly the station you want can be difficult this way since a tiny movement of the rod can change the tuning to a different station.

Finer tuning is done by turning the brass rod in the variable capacitor. To make this easier to do, and to make it easier to make fine adjustments, make a large knob out of a plastic lid from a jar or can that you no longer need for anything else.

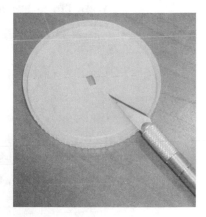

With a small, sharp knife, cut a small rectangle out of the center of the lid. The rectangle should be just a bit smaller than the brass rectangular top of the rod in the variable capacitor, so it will make a very tight fit when you press it onto the brass rod.

The photo below shows the tuning knob in place. The three pushpins hold up the knob so it doesn't wobble. With the large knob, it is easy to select the station you want to hear.

Since the ferrite rod is still loose in the coil, the radio is not yet very portable. At this point you need to find out where to place the rod so that all of the stations in the AM band can be

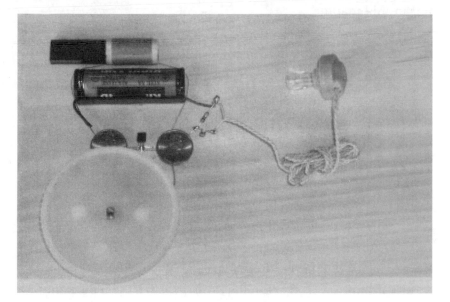

tuned using just the variable capacitor. You can do this by turning the capacitor all the way to the left and then sliding the ferrite rod into the coil until you hear the first station. Now you can tape the rod onto the board, or glue it there with some silicone rubber glue. You can also glue down the battery holder if you like.

Your Three-Penny Radio is now complete!

WHY DOES IT DO THAT?

If you have been reading from the beginning of the chapter, you probably already know most of the science behind how this radio receiver works, since it is very similar to a crystal radio.

Like a crystal radio, this is a "tuned radio frequency" receiver. That means it listens to the radio signal directly. It does not contain an oscillator like some other radio circuit designs, such as superheterodyne and regenerative radios.

The coil and variable capacitor join together to form a "tank circuit" that selects which radio station you want to listen to. Tank circuits and capacitors are covered in considerable detail and length in the "Adding a Capacitor (or Three)" section earlier in this chapter.

The main difference between this radio and a crystal radio is that the integrated circuit in this radio not only has the crystal inside it but has amplifiers and an automatic gain control.

The antenna coil (the little coil with the ferrite rod inside) generates a tiny amount of electricity as the radio waves wash over it. An amplifier is a circuit that uses that tiny amount of electricity to control a much larger flow of electricity from the battery. It is like using the water from a garden hose to move the nozzle of a firehose, putting a huge amount of water where you want it using only a little water from the garden hose.

The automatic gain control circuit controls how much amplification is used. It turns up the volume on weak stations so they sound as loud as strong stations do. This is why you don't need a volume control on your radio; all the stations are approximately the same loudness. (No AGC circuit is perfect, however; you can still tell which stations are powerful nearby stations and which ones are far away or weak).

In the radio shown in the photos, we use a 1.5-volt battery— in this case a small N cell, but you could also use a D, C, AA, or AAA cell.

The radio will work with battery voltages as low as 1.1 volts or as high as 1.8 volts. The current needed is very small—only 3 milliamps. This tiny amount of electricity is easily obtained from homemade batteries or small commercial solar cells.

One simple homemade battery can be made with just a piece of crumpled aluminum foil in a stainless steel bowl of vinegar and salt. The foil is kept from touching the bowl by a piece of paper tower or newspaper.

The stainless steel bowl and aluminum foil must not touch one another. You can get higher voltage by connecting the bowl of one battery to the aluminum foil of the other battery. This is a series connection.

The radio needs between 1.1 volts and 1.8 volts to operate. But it also needs at least 0.1 milliamps of current. The specifica-

tions say it needs 3 milliamps, but as you can see in the photo, it uses only 0.15 milliamps, and the radio has very nice volume.

The voltage is determined by how many bowls you have. The current is determined by how much surface area the bowls and aluminum foil have. Using bigger bowls and more foil will produce more current.

The bowl is the positive wire and connects to the radio where the red wire from the battery holder went. The aluminum foil is the negative side of the battery and connects where the black wire from the battery holder connected.

You can also try soft drinks or lemonade instead of the vinegar. Some people power their radios with beer. Depending on the beer, you may need more than three bowls. Adding salt to the beer will keep it from disappearing into curious bystanders.

Some Fun Packaging

The Three-Penny Radio is small enough to fit into some fun and interesting containers. I found a nice wooden box at a local store and built a radio to fit inside it. Instead of pennies, I used upholstery tacks, stuck into a bit of cork for a base. The cork was cut to fit the box.

The tuning knob is a plastic soda straw glued to the brass shaft of the variable capacitor and exiting out of a hole drilled in the back of the box. I used a small N cell battery that fit nicely in the box, and powered the radio for weeks (there was no off switch). You could remove the battery when not using the radio to make it last longer. The earphone coiled up inside the box for storage.

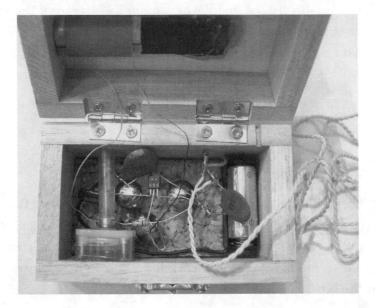

At another local store I found a little soap dish that was just begging to be turned into a radio.

I cut a slot in the lid to let the earphone wire come out while the lid was on, so the radio fit in a shirt pocket nicely with the lid on. Like in the wooden box radio, the earphone coiled up inside the box for storage.

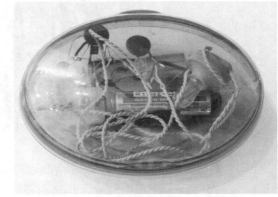

5

THERMODYNAMICS

Heat engines take many forms, from the internal combustion engine in the family car to the giant turbines that generate electricity for our homes. What they all have in common is that they steal some of the energy moving between something hot and something cold, and convert it into motion.

When you heat a gas like air or steam, the molecules in the gas move around faster. The faster they move, the harder they hit anything that is in the way. If something is in the way, such as a propeller or a pinwheel, the molecules can make it spin. This is how turbine generators spin to create electricity. If the gas is confined in a container with a lid, it can pop the lid off. This is how the engine in a car works. If fast-moving molecules push on one side of a container and escape through a small hole on the other side, so they are pushing on one side more than on the other, they can create a rocket or a jet, which moves away from the side with the hole.

In this chapter you will make a working version of the first heat engine (a form of steam turbine), a steamboat, a bimetal strip engine, and some tiny rockets that can shoot across a driveway.

Hero's Steam Engine

In the first century A.D., the mathematician Hero of Alexandria described a device called the aeolipile, in which steam was conducted through pipes from a boiler to a sphere that had two jets from which the steam could escape.

Hero's Steam Engine (if it was ever actually built) was probably a device that was not put to any useful purpose. Continuing in that fine tradition, you will build a version of his engine for the same purpose. Your version will be much simpler to build since you won't need steam-proof bearings for an axle.

SHOPPING LIST

- Can
- 2 thin brass tubes
- Very small nail
- Solder
- Beaded brass chain
- Sheet of rubber
- 3 pieces of wood
- 2 wooden dowels
- Small can of Sterno

TOOLS

- Diagonal wire cutters
- Pliers
- Soldering iron
- Scissors
- Drill
- Glue

Start with a can. The can will be both the boiler and the rotating part with the jets. This simplification eliminates any need for steam to enter the jet chamber through tight-fitting bearings. Insert two thin brass tubes into the can near the top, opposite one another. Bend the tubes so they are tangent to the can and crimp them almost shut so the steam pressure can build up inside the can.

I have made several aeolipiles from various tin cans, and some work better than others. Small cans, like the ones tomato sauce comes in, work well, but they have no convenient screw tops through which to add water. One must draw the water in through one tube by sucking on the other, which is not a very dignified way to show off first-century high technology. The best can I have found to date is a small can used to hold Oatey Purple Primer, a solvent for PVC plumbing pipe. This product is inexpensive and easy to find in most hardware stores. Pour out the contents into a jar and give it to your favorite plumber, then wash out the can well.

The Oatey can has a wire soldered to the lid and a ball of fibers at the other end of the wire to act as a brush for applying the primer. Cut the wire off as close to the lid as possible using diagonal wire cutters. Purists may unsolder the wire, but this takes longer. Dispose of the wire; you won't be needing it.

Next, use a very small nail to punch two holes near the top of the can, ⅛ inch from the top, so you have a little wiggle room. I usually use an old jeweler's screwdriver, but a small nail works fine. The holes must be directly across from one another or the engine will wobble too much. (It will wobble a lot anyway, so don't worry too much about precision.) Two small brass tubes, about one inch long and as thin as you can find, go into the holes, one tube in each hole. Brass tubing can be found in most hobby stores. It can be cut with diagonal cutters, but then one end must be filed down so that it is wide open. You can open the other end by squeezing it with a pair of pliers, since you don't want it open all the way. The tiny opening you get by squeezing makes a perfect nozzle for the steam to escape.

The wide-open end of the tube goes inside the can, and the nozzle end sticks out. Press the tubes against the can so they are tangent to the can, rather than sticking straight out.

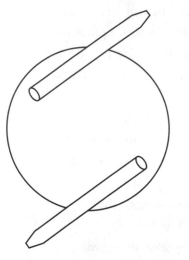

Now solder the tubes to the can. Use a high-wattage soldering gun or iron since the can will absorb a lot of heat. The little soldering pencils used for fine work in printed circuit boards aren't good for this kind of plumbing. (Radio Shack's 100-watt soldering gun works great.) Always use lead-free solder on any device or object that will be handled, especially by children, who are more sensitive to lead poisoning. Check that steam will be able to flow through the tubing by blowing into the can to see if air comes out the nozzles.

Next, solder a length of beaded brass chain, the kind used as a pull chain to turn lights on and off, to the top of the lid, in the center. This chain will act as a support to hold up the engine, and as a bearing to allow the engine to spin. I have tried fishing swivels, but the chain has less friction and is cheaper. I usually solder the connector of the chain to the lid, since it is easier to handle with a pair of long-nosed pliers. But if you are making more than one engine from one chain, simply solder one of the beads to the lid, taking care that the hole in the bead where it connects to the rest of the chain is straight up. The chain can be anywhere from 3 inches to 12 inches long, depending on how far up you want to put

the support (which will be described later).

Take the lid of the can and use it to trace a circle around a sheet of rubber, either a piece of an old inner tube or a bit from a tire patch kit. Cut out the circle and push it into the lid of the can as a seal—called a *gasket*—to keep the steam from coming out where it shouldn't.

The support is made of three pieces of wood. One piece forms the base; it can be any flat piece of wood, such as a two-by-four. Wood pieces that have been shaped for plaques work nicely. The base should be large enough so that the engine does not topple over as it wobbles around. A minimum size is about 3 by 5 inches.

Two wooden dowels form the rest of the support. One dowel should be ½ inch or more in diameter and about 15 inches long, to form the mast. Drill a hole in the base that fits the mast snugly. If the fit is not tight, you can use glue to hold the pieces together. The last piece is a small dowel about 5 inches long and ¼ inch in diameter. Drill a ¼-inch diameter hole in the mast about an inch from the top, and force the small dowel into it. The fit does not have to be snug, but a snug fit will keep the pieces together when the engine is trans-

ported. Do not glue this part, since it will be adjusted later.

You must also drill a hole in the small dowel about an inch from the end. This hole should be just a little larger than the beads in the beaded chain, since the chain will be threaded through this hole. If the fit is a little snug, you will not need anything to hold the chain in place. If it is loose, a lit-tle bit of wire can be wrapped around the chain above the hole to keep the chain from falling through when the can is full of water.

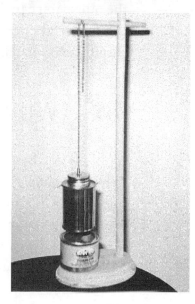

With the large dowel snug in the hole in the base, and the smaller dowel fitting in the hole near the top of the larger dowel, the chain now suspends the can a little above the base. Raise the chain enough to fit a small can of Sterno (jellied alcohol, used to keep food warm at buffets) under the Oatey can. Leave about ½ inch between the two cans so air can get to the alcohol.

Now fill the can with water, tighten the lid firmly, and light the Sterno to boil the water in the steam engine. If the can is full of water it will take a few minutes to boil. If the can has only a half inch or so of water, it will boil much sooner. Be careful not to let the can run dry or the can will become an ugly color, the solder will melt, and the rubber gasket will smoke and stink.

As the water boils, the can will start to spin. The steam coming out the brass tubes should be cool enough to touch (more on this later). If the steam does not exit with enough force to spin the can, there may be too much friction on the chain (due to too short a chain, not a brass chain, too heavy a can) but more likely the problem is with the nozzles. The nozzles must have very small openings so the steam pressure can build up in the can, making the steam exit the nozzles at high speed. Try pinching the nozzles shut a little more using a pair of pliers. Make sure there is a small hole in each, or the can will rupture from the pressure. I have built several of these and all of them worked the first time, although the ones with the smallest holes spun fastest.

As the can spins faster and faster, it will start to wobble. Even if you have built everything with the utmost precision, the water in the can is boiling and the bubbles make one side lighter than the other. You can ease the wobble by holding the chain near the can so the effective length of the chain is shorter. But be very careful not to touch anything hot.

WHY DOES IT DO THAT?

As the water heats up, the water molecules move faster. When the water boils, the molecules are moving too fast to stay stuck together as a liquid, and they move about freely as steam. The fast-moving water molecules bounce around in the can, hitting the walls of the can from all directions. Because they hit the top as often as they hit the bottom, the can neither moves up nor down.

But there is one direction in which the molecules don't hit anything. This is the direction where the holes in the tubes are. Instead of hitting a wall of the can, the molecules hit nothing and exit out into the air. The molecules in the can push on all the walls, except where the holes are. Since nothing is pushing in that direction, there is nothing to hold the can back, and it moves away from the holes in the tubes.

Imagine a big box on the floor, with no top, no bottom, and one wall missing. Let's call the walls the left wall, the right wall, and the front wall. The back wall is the one that is missing. Imagine 10 little kids are inside the box, all running in different directions. When a kid runs into a wall of the box the box moves a little bit, and the kid bounces off the wall and runs in another direction. Some kids will hit the left wall, and the box will move to the left. Some kids will hit the right wall, and the box will move to the right. These movements will cancel each other out, and the box will stay in the center of the room. Each time a kid hits the front wall, the box will move toward the front. But since there is no back wall, no kids will ever move the box backward. The result will be a box that moves across the room.

Rockets and jets move the same way the box moves. A rocket can work in outer space because it does not need to push against air or the ground. It works because the molecules inside the rocket are pushing in every direction except out the back. Hero's Steam Engine works because it has two rockets—the brass tubes—pushing the sides of the can in opposite directions, causing it to spin.

Why Is the Steam Cool When It Comes Out?

The steam is hot inside the can because of two things: the speed of the molecules, and the number of molecules crammed into that little space. The speed of the molecules creates the temperature, but temperature alone is not what burns your fingers. A lot of fast-moving molecules must hit your fingers before you can feel it as heat.

When the fast-moving molecules of steam leave the nozzles, they spread out in all directions as they run into the slow-moving air molecules outside the can. What you feel with your hand is the steam molecules that have slowed down in the air, and the air molecules they bumped into and caused to move. Not only have they slowed down by the time they hit your hand, but they have spread out so they don't all hit the same place. The combination of moving slowly and spreading out makes them feel cool on your hand.

The World's Simplest Steam-Powered Boat

This next device is the simplest example of a steam engine you will ever see. It has no valves and no moving parts (in the traditional sense of the phrase), yet it can propel its little boat easily across the largest swimming pool or quiet pond.

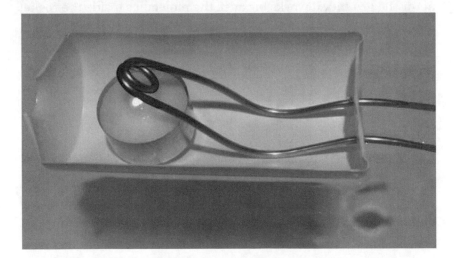

SHOPPING LIST

- Plastic bottle
- Small candle
- ⅛-inch soft copper tubing
- Nail

TOOLS

- ☐ Sharp utility knife
- ☐ Sticky tape
- ☐ Tubing cutter or hacksaw
- ☐ Sandpaper (optional)
- ☐ Large pen or pencil

The photograph on page 142 reveals nearly all the construction details.

The boat is a plastic bottle cut in half lengthwise. I used a soft plastic bottle that contact lens saline solution comes in. It is made of white high-density polyethylene (HDPE), and is very easy to cut using a sharp utility knife.

A small candle sits near the bow of the boat. I chose a candle that comes in a small aluminum cup, one used to keep food warm on the table. It is less than an inch high, so it sits low in the boat and won't drip wax all over. The aluminum cup keeps the melted wax in place, so the boat can go for hours. I also used a small amount of sticky tape to keep it in place.

The engine is made of ⅛-inch soft copper tubing. You can find the tubing at better hardware stores; it is normally used in refrigerators. You can use the easier-to-find ¼-inch soft copper tubing, but the ⅛ works best with the candle's small heat source, and because it weighs less it is less likely to tip the boat.

The best way to cut the tubing is with a tubing cutter. (You can get one at the hardware store when you get the tubing). You can also use a hacksaw, but you will need to clean the debris out of the cut with a knife or sandpaper.

Gently bend the tubing around a large pen or pencil—I used the handle of a wooden spoon—to form the coil in the center. Poke two holes in the back of the boat with a nail, and force the copper tubes through the holes. The holes in the soft plastic will close around the tubing, forming a watertight fit. You can also simply lay the tubing on the top of the back part of the boat and then bend it down so the ends are underwater, pointing to the rear. Carefully bend the tubing so the coil is just above the top of where the candle flame will be.

The boat is now finished, and ready to launch!

The copper tubes must be full of water, and both open ends must be underwater. The easy way to fill the tubes is to hold one end under water and suck on the other end. After the tubes are full of water and the boat is resting in the water with both ends of the tubing under the water, light the candle.

When the coil of copper tubing is hot enough to boil the water inside, the boat will jerk ahead suddenly, then start moving evenly forward. If you put your fingers in the water just behind the tubes, you can feel little pulses of water, about 5 or 10 pulses per second. These pulses are pushing the boat along.

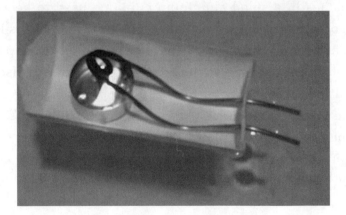

In the photograph on the preceding page, the candle has been replaced with a lump of charcoal starter cake. This makes a bigger flame, which makes the boat move even faster. The candle is safer, however.

In the photograph, you may be able to just barely see the ripples in the water behind the boat. These ripples are caused by the 5 to 10 pulses per second that the engine makes as it operates. The easiest way to view this when you make your boat is to watch the reflection of a bright light on the surface of the water.

WHY DOES IT DO THAT?

When the water in the coil boils, the steam expands. This pushes the water out of the tubes. The reaction pushes the boat forward, as explained above.

As the steam continues to expand, it encounters the section of tubing that was once full of water. This tubing is cold, and the steam condenses back into water. This causes a vacuum to form, which pulls more water back into the tubes.

You would expect that the water moving back into the tubing would cause the boat to go backward. However, the water doesn't get very far before it hits the end of the tube (the two streams of water in the two tubes meet each other in the coil). The water hitting the front of the tube (the coil) and pushing the boat forward again reverses any motion caused by the water being sucked into the tubes. As you saw when you put your finger near the tubes, this back-and-forth water motion is fairly rapid, and the comparatively heavy boat never actually moves backward at all.

A Rotary Steam Engine

You can use the same steam engine that pushes the boat in a different way, to cause something to turn.

SHOPPING LIST
- Aluminum soft drink can
- Aluminum foil
- Copper tubing
- Candle

TOOLS
- Scissors
- Pliers
- Paper punch

Start with an aluminum soft drink can. Cut it about a third of the way up from the bottom to form a small cup.

Carefully (so as not to cut your fingers) bend over the sharp edge, and crimp it with a pair of pliers to form a neat edge. Don't worry too much about neatness, since that will not affect the operation of the engine.

Put a large dent in the bottom of the can, the part that is normally bowed up in a dome; you will be putting a candle in the dent later. Punch two neat holes in the side of the can, opposite one another, with a paper punch. These holes will hold your copper tubing.

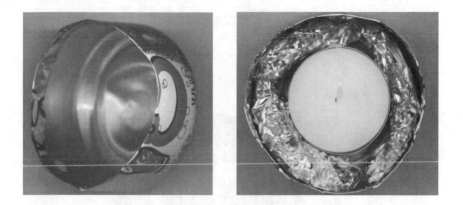

Next, put in the candle—the same kind of candle (with its convenient little cup to hold the melted wax) you used for the boat. Around the candle, put a rope of crumpled aluminum foil to keep the candle centered in the can.

Now make the tubing coil in the same way you did for the first boat. This time, however, bend the ends out to fit through the punched holes. This is a little tricky to do at first, but you soon get the hang of it. In the version shown, I have added an extra coil to the engine, but this is not necessary and does not improve the engine.

Next, bend the ends of the tubing gently so they curve into a right angle. Bend each end opposite the other, so the can will spin when it is placed in the water and the engine is started.

The final photograph shows the engine in operation. With a candle, it spins at about one rotation per second.

The Film Can Cannon

This device was an instant favorite from the moment its first loud bang and flash of orange flame launched the little black film can up to bounce off a 26-foot ceiling. It has several names: the Piezo-Popper, the Binaca Bomb, the Photo-Flash—you will probably come up with more.

SHOPPING LIST

- Perfume or aerosol hairspray or Binaca mouth freshener
- Pair of wires
- Plastic 35-millimeter film can
- Solder
- Igniter element from an electronic cigarette or fireplace lighter
- Block of wood

TOOLS

- Soldering iron

It is very easy to make, taking about 15 minutes, at a cost of two or three dollars if all the parts are purchased new (or free if you use certain common household items). The fuel for the cannon can be found around the house. I have run mine on perfume, aerosol hairspray, and (my favorite) wintergreen Binaca mouth freshener.

The cannon is very simple. A pair of wires is pushed through a slit made in the top of a plastic 35-millimeter film can. The other ends of the wires are soldered to the igniter element from an electronic cigarette or fireplace lighter. I chose to mount these elements on a block of wood, but this is optional if you're in a hurry.

Before firing the cannon, consider these safety tips. Do not give the cannon to anyone you would not trust with a box of matches. Adults should read these safety and operating instructions to the child before using the cannon. The cannon is loud. Do not shoot the gun close to the ears of people or pets. Do not fire the cannon in the direction of people or pets. Don't put anything inside the film can except the Binaca. Other things (such as paper, grass, or flammable liquids) can ruin the gizmo when they catch fire. Binaca is the most powerful propellant around; it works better than gasoline, lighter fluid, alcohol, or hairspray. It is also much easier to get the right fuel-to-air mixture with the measured spray of the Binaca can. Don't use more than one spray of Binaca. Too much fuel will prevent the explosion, but may not prevent the liquid residue from catching fire and ruining the cannon, or causing other serious damage. This cannon works like an automobile engine: it uses a spark, air, and a gas, all in the right proportions. Always keep the cannon clean, and never overload it. Follow the instructions and never misuse or abuse the gizmo.

That said, it's time to fire the cannon. Spray the fuel (one squirt of perfume or Binaca, or a very short squirt of hairspray) into the plastic film can, push the can down on the lid, and press the igniter.

Move your head away from the cannon. With a loud bang and a flash of orange flame, the little can will go sailing into the air. With a little practice in getting just the right amount of fuel in the film can, it will go as high as 30 feet. If too much or too little fuel is used, it will either not ignite at all or will not go very high.

The finished cannon is shown above, next to a bottle of perfume (the fuel).

The photo below shows the cannon with the can removed from its launch pad. You can see the two wires coming through the slit in the lid.

The block of wood has three holes drilled in it. Two go all the way through, so the wire can be threaded through them. The other hole does not go all the way through; it is used to hold the igniter.

The photo at right shows the bottom view, showing the wire threaded through the holes. You can also see the stick-on rubber feet that I attached to keep the wire from making the block wobble. You can get these feet at a cabinetry supply shop or hardware store.

Below (left) is a close-up of the spark gap formed by the stripped ends of the two wires. There is nothing critical about this arrangement; as long as the wire ends are bare and close enough, a spark can jump across the gap when the igniter is pushed.

Above (right) is a close-up of the igniter, showing the wires soldered to its contacts. There are many types of igniters in the different brands of electronic lighters.

Here is the igniter from a small Scripto electronic lighter. It is smaller and not as sturdy as the ones from the fireplace lighters, but the small lighters are usually less expensive.

At left is a small lighter disassembled to show the igniter. The lighters come apart easily without tools.

Finally, the photo below it shows the igniter from a large fireplace lighter. The larger igniters take more abuse than the little ones, and they are much easier to connect the wires to since they have little copper tabs that solder easily. They are designed so soldering isn't necessary, but I solder them anyway for a more permanent, solid connection.

WHY DOES IT DO THAT?

While perfume (which is mostly alcohol) works pretty well, the best fuels are aerosol hairspray and Binaca. Aerosol hairspray also has alcohol in it, but it also contains large amounts of propane, butane, and isobutane as propellants (gases under so much pressure that they are liquids in the can and turn to gas at the nozzle). These gases are excellent fuels. Binaca is a mixture of alcohol and isobutane, and comes in a convenient dispenser. It fits easily in a pocket and delivers just the right amount of fuel in a single push of its button. (Because a hairspray canister keeps spraying, it is harder to get just the right amount.)

In order to make an explosion, you need a flammable gas, oxygen, and a source of heat to start things off. Solids, like candle wax, and liquids, like alcohol, only burn when they are heated enough to become gases. Then they need a little more heat to break their chemical bonds so they can combine with the oxygen. By starting off with a gas, like propane, or a vapor, like that from alcohol that has been sprayed in a fine mist, you only need a small spark to start things burning.

The Film Can Cannon can holds only a small amount of air and fuel mixture, so it is safe to fire off in the house. The plastic can is soft and light, and can land on a person without disturbing his or her hairdo. But since it takes off rather quickly, you must keep your head out of the way during a launch to avoid injury.

The amount of air required to be mixed with the fuel will vary with the fuel used. The *fuel-air ratio* must be just right for some fuels. Other fuels, such as hydrogen, have a wide range of ratios that will explode.

Hydrogen will burn in air at concentrations ranging from 4 to 75 percent by volume. Methane (natural gas) burns at 5.3 to 15 percent, propane burns at 2.1 to 9.5 percent, and isobutane burns at 1.8 to 8.4 percent. Hydrogen will *explode* in air at ratios of 13 to 59 percent, while methane explodes within a much narrower range, between 6.3 and 14 percent (ratios are fuel to air).

It is easy to see how too little fuel will result in no explosion. But the problem is more likely to be too much fuel. If your can won't go bang, try lifting it off the pad and putting it back. This will allow a little more air in, and you will probably get a bang out of the results.

As the fuel-air mixture burns, energy is released by the formation of chemical bonds between the oxygen in the air and the carbon and hydrogen in the fuel. This energy heats up the gases that result from the burning. The gases are water vapor (H_2O) and carbon dioxide (CO_2). Since they are hot, they expand. The expansion pushes on all sides of the can and its lid. The can and the lid separate quickly, and the can goes skyward.

A more detailed explanation of exactly how the expanding gases cause the can to move can be found in the section on steam engines at the beginning of this chapter.

How Does the Igniter Work?

The igniter is a piezoelectric generator. The word *piezo* comes from the Greek work for "press." A piezoelectric substance is something that makes electricity when you press on it.

The classic example of a piezoelectric substance is a quartz crystal. Quartz is made up of atoms of silicon and atoms of oxygen. These atoms are arranged in neat, orderly rows. By carefully cutting the crystal, you can arrange for the rows of oxygen atoms and silicon atoms to be parallel to the cut surfaces, as in the following diagram:

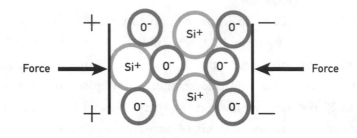

When pressure is applied to the crystal, the negatively charged oxygen atoms move relative to the positively charged silicon atoms. This causes electrons in the metal contacts to move, generating electricity.

The piezoelectric material in the igniter is not quartz but is a man-made ceramic that has been formed under a high-voltage electric field to align the electric charges in it. These manufactured ceramic piezoelectric materials can generate very high voltages.

The igniter holds this ceramic element in a plastic case with a steel hammer attached to a spring and a catch. As you push down the plunger, the spring is compressed until it hits the catch, which releases the spring, quickly pushing the hammer down on the ceramic. The electricity runs through the wires to the spark gap, which it jumps across, igniting the fuel-air mixture.

The Mark II Film Can Cannon

I found these little wooden spoked wheels at an upscale lumberyard. I just had to build a film can cannon with these wheels.

A little walnut, a little rosewood, a little clever drilling to hide the wiring, and we have the Mark II Film Can Cannon.

This cannon is angled at 45 degrees to maximize the distance traveled horizontally; however, due to wind resistance, that angle is not exactly optimal.

Of course, this cannon fires its entire barrel, but I can live with that. The cannon in the photo above is made from a kit, sold at www.scitoys.com.

A Bimetal Strip Heat Engine

This device is based on the simplest, most common object in the house that moves when heat is applied—a thermometer.

In this case, the thermometer is not the type with a liquid in a glass tube, but rather the dial type, where a needle moves like the hand of a clock across the numbers on the dial. The heart of this type of thermometer is something called a bimetal strip, which is simply two strips of different metals, bonded together. In the thermometer, the strip is wound up into a spiral and the dial is attached to the center.

The concept behind the Bimetal Strip Heat Engine is very simple.

Suppose you place a dial thermometer on a small teeter-totter. Then you put a weight on the end of the dial and put another weight on the other end of the teeter-totter to almost balance it.

If the thermometer is cold, the weight will lean toward the end of the teeter-totter (where the low numbers are on the dial). When this happens, the teeter-totter will be unbalanced and the thermometer side will hit the ground.

If the ground is hot, the dial will turn toward the higher numbers, carrying the weight toward the center of the teeter-totter.

This shift in the balance will cause the teeter-totter to tip the other way, raising the thermometer off the hot ground and up into the cooler air.

As the thermometer cools, the weight shifts again and the whole process repeats itself over and over again.

In this heat engine, you will use only the most important part of the thermometer, the bimetal strip. The first engine shown here uses a large coil from a big thermometer I found at a local hardware store. The face of the dial was 12 inches across. The second engine is built from a smaller thermometer.

SHOPPING LIST
- Bimetal strip
- Coil
- Brass screw
- 1 or 2 nuts
- 2 small pieces of copper wire
- Solder
- 2 coat hangers
- Block of wood
- Cup of very hot water

TOOLS
- Pliers
- Soldering iron
- Sandpaper
- Drill

The bimetal strip usually has one edge bent away from the coil to hold the coil in place in the thermometer. This makes a nice place to attach the weight.

In these engines, the weight is a brass screw with a nut or two on it. The nut allows you to adjust the balance of the weight to make the engine run its best. You can move the nut up or down by simply turning it, which allows much finer adjustments than you actually need.

The screw is attached to the coil by two little pieces of copper wire, twisted tightly with a pair of pliers. (While not necessary, I removed the head from the screw with a large pair of wire cutters since the head wasn't needed.)

The center of the coil is attached to the teeter-totter with more copper wire. In this case, I put two pieces of wire, one on either side of the inside end of the strip, against the strip and twisted them tightly with pliers. I then soldered the twisted copper wires to hold them tight.

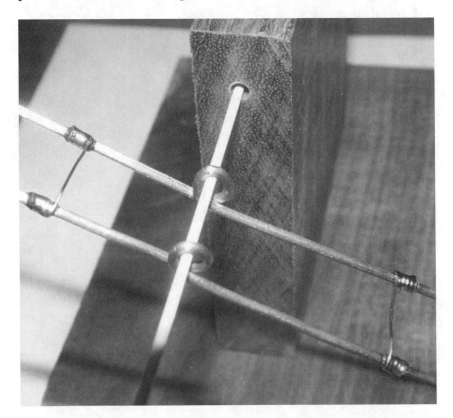

The teeter-totter is made from two coat hangers. Sand the paint off the coat hanger wires so they can be soldered, and bend the wires around a nail in the center to form a pivot.

Solder the wires from the center of the coil to the two coat hanger wires to complete the coil end of the teeter-totter. The weight should be pointing all the way to the outside of the teeter-totter, away from the pivot, when the thermometer is at room temperature. If necessary, put a couple of copper wires across the coat hanger wires between the pivot and the ends to make the assembly a little sturdier.

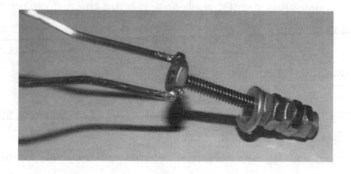

The counterweight is made with another brass screw. In the first engine, it was attached the hard way—by soldering a nut to the coat hanger wires and screwing the screw onto the nut. In this engine, the head of the screw is attached to the coat hangers with some copper wire, then the coat hangers were soldered to the wire. The nuts can then be easily screwed onto the other side of the screw. I recommend this approach; it's much easier.

The pivot is simply a nail driven into a block of wood. In these engines, I drilled holes in the wood and pushed pieces of coat hanger wire into them, but a nail will work fine.

One thing worth noting at this point: the coat hanger wires are bent slightly upward, so that the pivot is below the center of gravity of the teeter-totter. This means that if you held the whole thing up by the pivot, it would want to turn upside down. It is supposed to be unstable, the opposite of the kind of balance used to weight things. Keep it from turning upside down by holding it low, close to the table. This way one end or the other will usually be touching the table.

Operating the Engine

To operate the engine, pour a cup of very hot water (or very hot tea), and set the coil in the cup. It should quickly flip the weight toward the center of the teeter-totter. If it doesn't flip the weight, the weight may be too heavy or may be located too far out on the screw. Make adjustments, such as screwing the nut closer to the coil or removing a nut or two. You want as much weight as the coil can easily flip, placed as far from the coil as possible, while still allowing the coil to flip the weight.

You may find it useful to put something under the counter-weight to keep it from going down too far. The coil is a spring, and when the teeter-totter brings the coil way up, the weight pushes on the spring so hard that the coil may not be able to raise it back up. You can keep this from happening by not letting the coil rise very far.

Some Improvements in a Smaller Engine

To make sure that less expensive thermometers will work as the basis for the engine, I built a second engine using a smaller bimetal coil. At the same time, I incorporated some improvements I learned from working with the first engine.

The second engine has a smaller teeter-totter. This is not because the coil is smaller. Instead, it is to make the movement of the weight have a greater effect on the balance of the teeter-totter.

Ideally, a long plastic screw would probably be better than the brass screw as a holder for the weight. This would allow the screw to be long enough to hit the pivot (or even farther), creat-

ing a much bigger unbalancing effect on the teeter-totter. However, I used what I had on hand.

Making the teeter-totter smaller allowed me to make it out of a single piece of coat hanger wire. This made the construction much easier and the resulting assembly much sturdier.

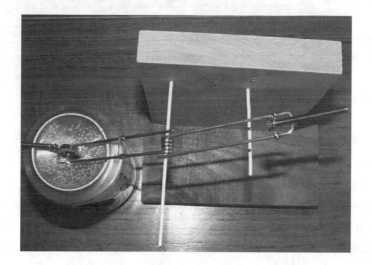

I added a piece of copper wire to prevent the weight from traveling past the teeter-totter wires. I had found that if the water was very hot, sometimes the weight swung all the way around and stopped the engine from working.

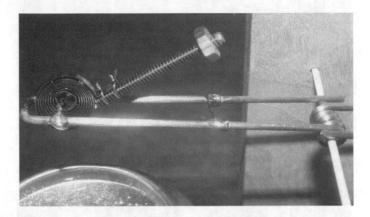

As I already mentioned, I also found a much easier way to attach the counterbalance screw.

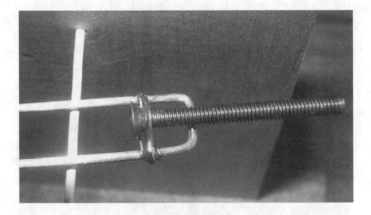

Lastly, I added another wire coming from the block of wood to act as a stop for the teeter-totter so it would not tip so far that the weight got stuck in the inside position.

I also improved on the cup of tea. I found that as the tea cooled, I had to keep adjusting the counterbalance weight to keep the engine working. In the new engine, I keep the water boiling by using a cut soda can and a candle. The water sits in the dent in the bottom of the upside-down soda can.

🤖 WHY DOES IT DO THAT? 🤖

There are two important principles in the operation of this engine. One is *differential expansion* and the other is *hysteresis*.

Most metals (and many other materials) expand when taken from room temperature to the temperature of boiling water. But each metal expands at a different rate. The bimetal strip in the thermometer is made up of two metals that have very different expansion rates.

In the coil, the metal on the outside expands more quickly than the metal on the inside. This causes the coil to curl up when it is heated. The heat energy in the hot water is converted into mechanical energy as the strip curls up. This mechanical energy is used to put the weight closer to the center of the teeter-totter, which is a change in potential energy. This potential energy is converted back into mechanical energy as the teeter-totter rocks away from the hot water.

The second principle, hysteresis, is what keeps the engine from simply rising to a certain point and stopping, like a normal thermometer would do.

As I mentioned earlier, the coil acts like a spring. This allows the weight to act against the curling of the coil when the coil is low, and then suddenly act with the curling coil when the coil is high. The result is that the teeter-totter is more stable with one or the other end down, and is unstable in the in-between position. This tendency to quickly flip from one state to another is hysteresis. The engine is an example of a *bistable system*, sometimes known as an *astable oscillator*.

A Metal That
Melts in Hot Water

Take a look at the photo below. As the silvery metal wire is held in the steaming hot water, drops of liquid metal form and fall into a shining puddle at the bottom of the glass. What was once a thick, solid wire is now a pool of molten metal. No soldering irons, no

flames or torches, just a cup of hot water.

Since this liquid metal is only 144°F (62°C), it can be poured into plastic molds designed for making candles or soap.

In the picture below, about a foot of the wire sits next to a ruler. The wire comes in a thick vinyl tube, so it looks bigger in the picture than its actual ⅛-inch thickness. A few inches were melted and poured into the plastic mold. A white plastic-coated paper clip was inserted into the mold before the

metal solidified, just for fun.

The metal is very good at filling all the nooks in a mold, to faithfully reproduce items as fine as fingerprints. On the facing page are some of the items we cast in metal, using only hot water.

What Is That Stuff?

This unique metal is a mixture of bismuth, tin, and indium. It is called Field's Metal. Indium is a metal that looks like silver but is about twice as expensive. It has special uses in scientific apparatus due to its interesting properties, such as its ability to stick to glass. This allows you to "solder" things directly onto glass, and to make a good metal-to-glass seal for vacuum work.

There are many alloys that melt at low temperatures. These are called *fusible alloys*. You may have heard of a famous one, called Wood's Metal. Wood's Metal is a mixture of 50 percent bismuth, 25 percent lead, 12.5 percent tin, and 12.5 percent cadmium. It melts at a temperature of 158°F (70°C).

Cadmium and lead are toxic metals. By using nontoxic metals like indium, tin, and bismuth, Field's Metal is safe to handle. It melts at a lower temperature as well. Indium is not a household material and can be hard to find. Commercial suppliers usually have minimum orders in the range of 10 pounds or more, bringing the minimum cost to several hundreds of dollars. For that reason, I have made up a batch of Field's Metal and cast it into wires, and made it available to experimenters at www.scitoys.com in affordable quantities.

Fusible alloys are used in some fairly common items. They are used in fire sprinkler systems in office buildings, where the metal melts if the temperature gets too high and lets the water spray out. They are also used in the pop-up timers for turkeys. A bit of solid metal holds a plastic flag down against a spring. When the temperature is just right, the metal melts and the spring pops the flag up into view.

⫶̣ WHY DOES IT DO THAT? ⫶̣

The melting point of an alloy is often quite different from the melting points of the pure metals from which it is made. Bismuth melts at 519.8°F (271.3°C). Tin melts at 447.8°F (231.8°C). Lead melts at 620.6°F (327.5°C). Indium melts at 312.8°F (156.6°C).

Pure substances have a sharp melting point. A pure substance is either an element or a chemical compound. Mixtures melt over a range of temperatures.

A eutectic alloy is one that has a sharp melting point. This implies that it is a chemical compound, where the elements are bound together in strict proportions, rather than a simple mixture of elements.

In a mixture of elements, some of them will react together to make compounds. Compounds consist of exact proportions of one atom to another, such as one to one, two to one, three to two, and so on. Any excess of one element over another will not react and will stay in the mixture as a pure element. That is why mixtures have a wider melting point. One chemical melts at one temperature, and the others melt at higher temperatures. Only when all of them have melted do you get a true liquid.

A substance will melt at a given temperature based on how strongly the molecules of the material bind to one another. Sometimes two elements combine to form a compound that binds tightly to itself. This would raise the melting point. Other times, the compound formed does not bind to itself as easily as the pure elements do. This compound would have a lower melting point than either of the pure elements. Compounds are not limited to two elements; sometimes many elements bind together into a single compound.

In electronics, one of the preferred solders used is a eutectic mix of lead and tin. By weight, it is about 63 percent tin and 37 percent lead. It melts at 361°F (183°C).

The atomic weight of lead is 207.2. The atomic weight of tin is 118.71. If there were one atom of lead for every three atoms of tin, the ratios would be:

$$\frac{207.2}{207.2 + 3 \times 118.71} = 36.78\% \text{ lead}$$

and

$$\frac{3 \times 118.71}{207.2 + 3 \times 118.71} = 63.22\% \text{ tin}$$

Thus the chemical formula for the compound is $PbSn_3$.

Another eutectic alloy is a mixture of lead and antimony. For every antimony atom, there are four lead atoms. The melting points are: Lead (Pb): 327°C, Antimony (Sb): 630°C, Alloy ($SbPb_4$): 246°C.

In the eutectic alloy of magnesium and lead, there are two atoms of magnesium for each atom of lead: Lead (Pb): 327°C, Magnesium (Mg): 651°C, Alloy ($PbMg_2$): 530°C.

Make Your Own Fusible Alloy

An easy fusible alloy can be made from bismuth, lead, and tin. Lead is toxic, so this alloy should be handled with care and should not be used near food or near children. This gizmo should definitely not be considered a toy. (For toy purposes, use Field's Metal instead.) It is important that you know how to melt solder if you plan to make your own low-melting-point alloy. If you don't have that skill, you are best off purchasing the alloy premade.

Bismuth can be found at sporting goods stores in the form of shot for use in shotguns. It is preferred to lead shot because it is nontoxic and doesn't pollute water.

Lead and tin are not hard to find separately, but they are particularly easy to find together in the form of solder. Solder is usually made in the eutectic proportions or in the nearly eutectic form of 60 percent tin and 40 percent lead. But some solder is available in the opposite ratio: 60 percent lead and 40 percent tin. This is the form you need (unless you can obtain your lead and tin as separate items).

The eutectic form of the bismuth-lead-tin alloy is 52.53 percent bismuth, 32.55 percent lead, and 14.92 percent tin, by weight. The compound is $Bi_8Pb_5Sn_4$. If you have the separate metals, you can weigh them out and melt them together, and you will have an alloy that melts at 203°F (95°C).

If you have the 40/60 tin/lead solder, you can weigh out equal parts of the solder and the bismuth and melt them together. This gives you a mixture that is not eutectic, so the melting point is a range from 203°F (95°C) to 219°F (103.8°C). This will still melt in near boiling water.

LIGHT AND OPTICS

A Simple Laser Communicator

How would you like to talk over a laser beam? In about 15 minutes, you can set up your own laser communication system, using cheap laser pen pointers and a few parts from Radio Shack.

SHOPPING LIST

Transmitter

- ◗ Laser pen pointer
- ◗ Battery holder that holds the same number of batteries as the laser pointer, often 3 cells (The batteries can be any size, but they must be the same voltage as the laser batteries. You may need to get one that holds two cells and another that holds one cell, and wire them together in series. Radio Shack has a good selection.)

- ○ Transistor radio (Later you will use a microphone and an amplifier [RS #33-1067 and #277-1008], but at first you will send your favorite radio station over the laser beam.)
- ○ Earphone jack that will fit your transistor radio [RS #42-2434]
- ○ Audio output transformer (It consists of an 8-ohm coil and a 1000-ohm coil. The one I used is the Radio Shack #273-1380. You can also get them at www.scitoys.com.)
- ○ Some clip leads—wires with alligator clips on the ends (At least one of the clip leads should be the type with a long slender point [RS #270-334] to connect to the inside of the laser pointer. You can substitute regular wire and solder if you like, but the clip leads are fast and simple. Radio Shack has a wide selection of clip leads, such as CRP #270-378.)

SHOPPING LIST
Receiver
- ○ Small solar cell, such as Radio Shack #276-124 (You may have to solder your own wires to it if it doesn't come with wires attached.)
- ○ Microphone jack that will fit the phono input of your stereo (RS #42-2434 or #42-2457) (Instead of a stereo, you can use the small amplifiers that Radio Shack sells [RS#277-1008])

TOOLS
- ☐ Rubber band or wire

First, remove any batteries from the laser.

Connect a clip lead to the inside of the laser pointer where the battery touched. Usually there is a small spring to which you can attach the clip lead. The other end of the battery usually connects to the case of the laser. Since there are many different styles of laser pointer, you may have to experiment with clip lead placement to get the laser to work with the new external battery pack. You may also have to hold down the laser's push button switch by wrapping a rubber band or some wire around it. Test the connection before you attach the resistors, to make sure the laser works with the new battery pack. If it doesn't light, try reversing the battery. Battery reversal will not harm the laser.

Connect the 1,000-ohm side of the transformer between the battery and the laser. The 1,000-ohm side of the transformer has three wires coming from it. Only use the outside two wires. The inside wire is called a center tap and you do not use it in this circuit. Test the laser by attaching the battery. The laser should operate normally at this point.

Connect the earphone jack to the laser, in parallel with the battery/resistor combination. The schematic of the transmitter looks like this:

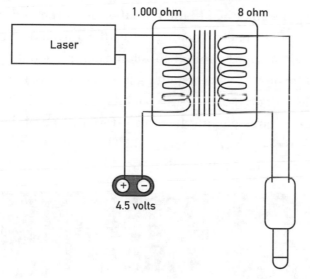

The transformer modulates the power going to the laser. The signal from the radio is added to and subtracted from the battery power, and the laser gets brighter and dimmer along with the volume of the music or voice in the signal.

The receiver is the simplest part. Just connect the solar cell to the microphone jack and plug it into the amplifier or stereo phono input. It does not matter which way the wires are connected to the solar cell. Here is the schematic of the receiver:

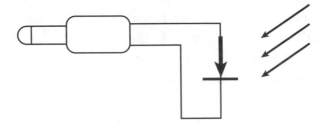

Setup and Testing

Make sure the transistor radio is turned off and the laser is on. Plug the earphone jack of the laser into the earphone socket of the radio.

Connect the solar cell to the amplifier or stereo and turn the volume up until you hear a hissing noise, then turn it down slightly until the hiss isn't noticeable. The volume control should be fairly high, corresponding to an earsplitting level if it were playing music.

Aim the laser across the room so it hits the solar cell. You might hear clicks or pops coming from the stereo or amplifier as the laser beam passes over the solar cell. This indicates that everything is working fine at this point.

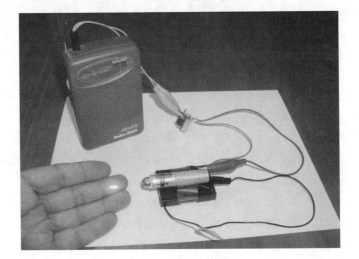

Now carefully turn on the radio and slowly adjust the volume until you hear the radio station voices or music coming from the amplifier across the room. Do not turn the volume up very far; it may damage the laser. The radio should be just audible if the earphone jack is pulled out, not loud. If you can't hear the sound from the amplifier across the room, make sure the laser is shining on the solar cell, then try increasing the volume of the amplifier before you increase the volume of the radio.

At this point you should be hearing the radio station coming in loud and clear in the amplifier across the room. Put your hand in front of the laser beam to break the connection, and notice that the music stops. Wiggle your fingers in the beam and listen to the music get "chopped up" by your fingers. Your laser communicator is ready for the next step.

To send your voice over the laser beam, simply replace the transistor radio with a microphone and amplifier. Radio Shack sells small amplifiers (about the same size as the transistor radio) that have sockets for microphones and earphones. You can also use another stereo system, but be very careful with the volume control to prevent damage to the laser.

Using a Disassembled Laser Pointer

For this project, remove the laser assembly from a small $10 laser pointer. The power supply circuit is the green board attached to the brass laser head.

The power supply circuit comes conveniently marked with a plus and a minus next to two holes in the board. Solder the black negative lead from the battery clip to the hole marked minus. Solder one of the 1,000-ohm coil leads to the hole marked plus. Solder the red positive lead of the battery clip to the other lead from the 1,000-ohm coil.

Attach the battery clip to a 4.5-volt battery pack (not a 9-volt

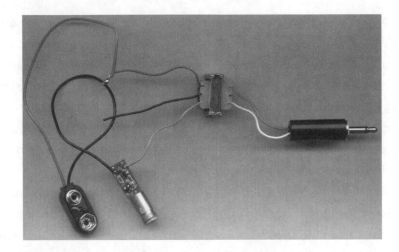

battery!). Since I didn't have a pack that takes 3 cells, I used one that takes 4 AA batteries, and I replaced one of the four batteries with a straight piece of bare wire.

That's it! You now have a laser transmitter!

A New Receiver

The solar cell receiver has some drawbacks. It is expensive—solar cells are a few dollars each—and fragile. A cheaper, stur-

dier alternative to the silicon photocell is a cadmium sulphide photoresistor.

A cadmium sulfide photoresistor is shown at left (magnified many times). It does not produce electricity from light the way a solar cell does. Instead, the light that falls on it changes its resistance to electricity. If you connect a battery and a photoresistor together, they can act like the solar cell. As the intensity of the light changes, the amount of electricity output changes in response.

The new receiver is very simple and looks like this:

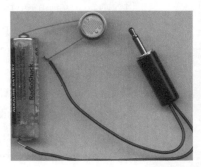

Super Simple Receivers

Using a supersensitive piezoelectric earphone, you can make a laser voice receiver that doesn't need any expensive amplifiers or power source. Just connect it to a small solar cell, as shown below:

If a solar cell is too expensive or fragile, a cadmium-sulfide photoresistor can also be used. The earphone wires are connected across the photoresistor, and the battery is also connected across the same wires. The battery, the earphone, and the photoresistor are in parallel. A 220-ohm resistor is placed in series with the battery to reduce power consumption and prevent heating of the photoresistor.

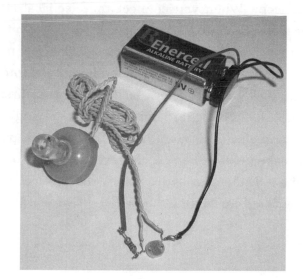

Either of these earphone approaches has the nice feature of making the communication private. Only you can hear what is coming over the secret laser link.

ᴦ⸱⎚ WHY DOES IT DO THAT? ⎚⸱ᴦ

In all of these laser communicators, the laser light is amplitude modulated. This simply means that the amount of light the laser emits varies over time.

To understand what is going on, it helps to consider how a loudspeaker makes sound. A loudspeaker is a paper cone attached to a coil of wire that sits in a magnetic field from a strong permanent magnet.

When an electric current flows in the loudspeaker coil, the coil becomes an electromagnet, and it moves toward or away from the permanent magnet. As it moves, the paper cone pushes on the air around it, compressing the air in front of it and expanding the air behind it. Waves of compressed and expanded air travel to your ear and cause your eardrum to move in time to the movements of the paper cone.

The laser communicator adds two components to the loudspeaker concept. When you connect the electrical signal that goes to the loudspeaker to the laser instead, the laser gets brighter and dimmer as the electric current varies. The second component is the receiver, which converts the light back into an electric current. This current varies in time with the first current because the amount of light that it receives is varying in time.

This second electric current is used to move the paper cone of a loudspeaker, just as before. However, now the loudspeaker can be quite a distance away from the original electric current, without any wires connecting the two.

Make Your Own 3D Pictures in Minutes

In this section you will see just how easy it is to take pictures that show realistic three-dimensional (3D) images.

The pictures can be viewed in three ways: by crossing your eyes, by focusing your eyes at infinity (called the "parallel" method because the two lines of sight are parallel), and with an inexpensive (or homemade) 3D viewer. The viewer is nice because it takes a little practice to see the images with the first two methods, and most people find the viewer easier and more comfortable.

Taking the Pictures

This is the simplest part. You can use any camera you have available. Just take a picture, then move the camera to the side a little bit, and take another picture. That's all there is to it! I like to use a tripod, but some people just shift their balance from one foot to the other for each shot.

If you have an instant camera, the pictures can (of course) be viewed right away. I like to use a digital camera because the pictures are of a higher quality, and I can still see them right away on the screen. Even if you use a standard film camera, the pictures can be digitized on a scanner and then pasted together to be viewed on the screen or printed on a color printer.

The next step is to place the two pictures next to one another and cross your eyes to see the 3D view. Place the picture that was taken from the right side on the left, and place the picture taken from the left side on the right.

To view cross-eyed, keep the pictures at a distance where you can comfortably focus on them. Slowly cross your eyes until you see three pictures instead of two. The center picture will be in 3D. It takes some practice. If you find yourself straining your eye muscles, you may be trying to focus on the air between you and the pictures, where your eyes are aiming. Relax and try again, letting your eyes focus on the pictures, but cross so the left eye sees the right picture, and the right eye sees the left.

Once you get the hang of it, you can do it comfortably right away and can view the pictures as long as you like, shifting your gaze from items in the foreground to items in the background effortlessly. The 3D effect is stunning, not only because of the stereo effect, but because there is twice as much information getting to your brain.

Sometimes it helps to start farther away from the pictures, and move closer only when your eyes are properly positioned and you can see the 3D effect.

Cross your eyes to view these images in stereo 3D.

My house.

Some people find it easier to aim their eyes at infinity rather than to cross them. Because two light rays coming from infinity are parallel when they reach your eyes, this method is called the "parallel" method.

The problem with the parallel method is that the pictures must be the width of the distance between your eyes. That's not very big, and that limits the detail you can get on a computer

The view from my home office window.

monitor. It is less of a problem with photographic prints (since they contain a lot more detail per square inch than a computer monitor).

For parallel viewing, the pictures are reversed from cross-eyed viewing. The picture taken from the right is placed on the right and is viewed by the right eye. The picture taken from the left is placed on the left.

Looking at the photos on the next page, relax and let your eyes drift through the pictures as if they were viewing a mountaintop far in the distance. You will gradually be able to see three pictures as with the cross-eyed method, and like then, the center picture will be in 3D.

For me, it sometimes helps to get very close to the image, so the pictures are very blurry. This makes it so that each eye is looking at a different picture. Then, when I have a blurry 3D image, I slowly back away from the image until it comes in focus, being careful not to lose the 3D sensation. When you are very close to the image, it will look like the two pictures have merged into one. As you back away, you will be able to see the other two pictures flanking the 3D center image.

View these pictures with eyes parallel (looking at infinity).

My house.

A path to the treehouse.

All of these photos have been done with the "hyper-stereo" technique, where the camera positions are separated by more than the distance between the eyes. For true stereo, try holding your head completely still (rest it against a wall for example), and hold the camera up to one eye for the first shot, then up to the other eye for the second shot. These 3D images will work well for objects that are nearby, and will not give the exaggerated 3D for distant objects that you see in the images of the lake in the photos above.

For very close-up objects, you can move the camera by less than the distance between your eyes. Now, instead of seeing the

3D effect behind the plane of the picture, the image seems to float in midair between you and the screen or paper. For this to work well, you may need a special close-up lens on your camera. However, sometimes just shooting the picture with a magnifying glass taped over the camera sunshade will produce very good results.

If you do want to play with the hyper-stereo effect, remember that the brain finds it easiest to see 3D images if the distance from the camera to the object is 30 times the distance between the two camera positions. If an object is 30 feet away, the camera position for the second shot should not be more than 1 foot from the position of the first shot.

The following pictures are done in true stereo, holding the camera first to one eye then to the other. The subject is the treehouse bridge between two trees in my yard.

Cross your eyes to view these images in stereo 3D.

The treehouse bridge.

View the next two pictures with eyes parallel (looking at infinity).

The treehouse bridge.

The treehouse bridge from the other side.

Remember, for the parallel viewing, each half of the picture must be about as wide as the distance between your eyes. A little less wide is usually OK, but wider won't work.

Using a Viewer to See the Pictures

There is an inexpensive viewer available from a company called 3-D ViewMax that makes viewing these images very easy and comfortable. The viewer is a simple pair of plastic prisms (with a bit of magnification) in a folding cardboard holder that keeps the pictures at the proper distance. The prisms make it easier for your eyes to view parallel format stereograms.

You can use the viewer to view pictures directly from the computer screen, or you can place your own pictures next to one another in the viewer. You can print yours on a good color printer on premium paper. The trick is simply to tell the printer to print the stereogram so it is 6 inches wide; the 3-D ViewMax viewer is designed for 6-inch pictures.

If you print your own photographs, make them 3 inches wide so the stereogram will be 6 inches wide when they are side by side.

It is a little difficult to build this kind of viewer yourself. The magnification is not strictly necessary so the viewer can be made by putting a small wedge prism in front of each eye. These prisms can be made by sanding and polishing small pieces of clear plastic, but this takes some skill. Another way to make a viewer is with small circular mirrors. The mirrors should be oriented 90 degrees from each other and separated by the same distance as your eyes. When you look into the mirrors, the left eye will be looking left and the right eye will be looking right.

The pictures are not stuck together in this viewer but are placed near the operator's shoulders, the left view on the left and the right view on the right. Such a viewer is more cumbersome to use than the 3-D ViewMax, but it is easier to explain how it works to a younger child, and, it makes a better science fair project.

Make a Permanent Rainbow

You have probably seen the beautiful rainbow colors caused by a tiny bit of oil floating in a puddle of water. In this project, you are going to capture those colorful patterns on paper in a permanent form so you can view them anytime you like, without a messy puddle.

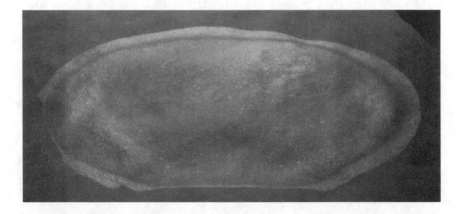

TOOLS

☐ Piece of paper (Black construction paper works well, but so does the coated glossy photographic paper used in ink-jet printers.)

☐ Cake pan or similar container that can hold water and is big enough to hold the piece of paper flat in the bottom (You can cut the paper to size if you don't have a pan big enough.)

☐ Clear fingernail polish

☐ Eyedropper (Warning: it will not be good for anything else when you are done. A disposable plastic pipette is ideal.)

☐ Newspaper

Place the paper in the bottom of the pan.

Fill the pan with enough water to cover the paper with at least half an inch of water.

Put the eyedropper into the nail polish bottle and squeeze just a little air out of it to get half an inch or so of nail polish into the eyedropper. You only need a single drop.

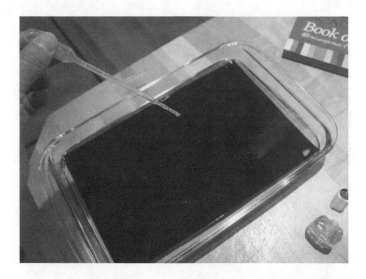

Drop a single drop of nail polish onto the surface of the water.

The nail polish drop will quickly expand to make a circle of film on the water. This film will be so thin that it will only be as thick as one wavelength of light.

Let the circle of nail polish film dry for a few minutes. The edges will generally wrinkle a bit, and the center will take longer to dry than the edges.

Gently lift one end of the paper out of the water, making sure you catch the edge of the thin circle onto the paper.

Let the water drip off the paper into the pan for a little while, and then set the paper (with the circle of film clinging to the middle of it) onto some newspaper to dry.

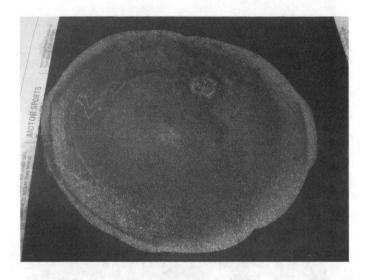

To view the colors, hold the paper flat toward the light and view it at a low angle.

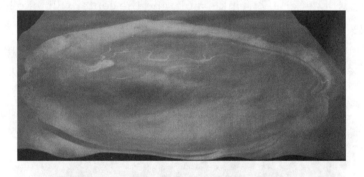

If you use more than one drop, you might get double patterns.

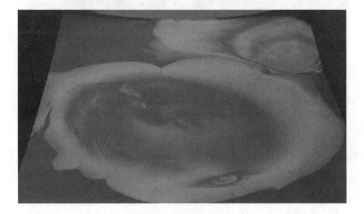

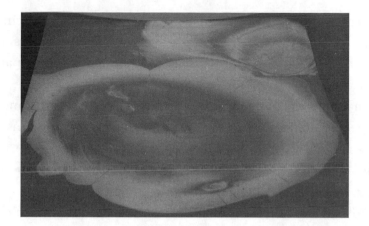

The glossy photo paper leaves a nice smooth surface with many rings of color.

🤖 WHY DOES IT DO THAT? 🤖

The colors on your Permanent Rainbow are caused by the interactions of several interesting qualities of light.

Light travels at different speeds in different materials. In air, or the vacuum of space, it travels very fast, about 299,792,458 meters per second. In water, light travels ⅓ slower than it does in air. Scientists say water has a refractive index of 1.33. In the film you made from the drop of nail polish, the light is even slower since the dried nail polish has a refractive index of 1.42.

Light travels in waves. You can imagine the waves of light as looking like waves of water in the ocean. The waves come by in parallel rows, one behind the other. Picture a set of waves coming in at an angle and encountering a material that slows them down, as in the diagram below:

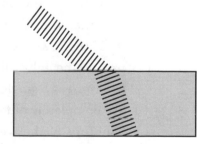

The left edges of the waves encounter the material first and slow down. The point just to the right travels a bit farther at the higher speed, but then it encounters the material and slows down. This happens all along the wave. Inside the material, the left part of the wave has been moving slower than the right side for a little longer as you look at each part of the wave.

Imagine rows of people, all marching along holding hands. If the people on the left start marching slower, the columns will make a small turn to the left. The waves of light do the same thing. When light encounters a material with a higher refractive index, it bends into the material.

But not all of the light bends in. Some of the light is reflected off the top surface. This is why you see sunlight reflecting off the top of the ocean. The light reflects off the top because of the difference in refractive index. But there is another place where the refractive index changes suddenly. At the bottom of the film, the light leaves the film and goes back into the air. But not all of it goes back. Some of it is reflected back up, just like at the top surface.

Light that reflects off of a top surface (from low index to high index) is completely out of phase with the light that reflects off the bottom surface (high index to low index). The rows of waves don't match up—it is like the rows of marchers missed a step.

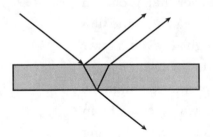

What you end up with are three rays of light. Two go up and one goes through the film.

The two top rays are probably not in phase. Their rows of marchers don't line up in matching rows. This is partly because the ray to the right has been shifted by the reflection at the bottom of the film. But it has also traveled farther than the other ray.

If the rows happen to be in phase, the marchers will all line up and the light will constructively interfere. It will appear twice as bright as a single ray since both rays are combined. If the rows are completely out of phase, the light will destructively interfere. The tops of the waves of one ray will fill in the troughs of the second ray, and you have no waves. No waves means no light, and the rays disappear. Somewhere in between those two extremes you will get some light, but not as much as two rays' worth.

Note that there are two things you can do to adjust the phase of the rays, to make them either constructively or destructively interfere at our command. The first thing you can do is change the angle at which the light hits the film. If the light comes in at a shallower angle, then the ray that enters the film will travel in a longer path in the film and therefore be slowed down more than before, and coming out the top a little later. The other thing you can do is to change the thickness of the film. If the film is thicker, then the ray that travels inside the film has farther to go and again comes out of the top a little later.

White light is composed of waves that have many different wavelengths. Different wavelengths are seen as different colors by our eyes. Different wavelengths also travel at different speeds in a refractive material like the film. This means that the ray on the right will be delayed more in some colors than in others. So while one color may experience constructive interference and be bright, other colors will be dimmer or completely gone.

Now you have all the pieces you need to understand the color patterns in the film.

At the outside edges, the film is thinnest. The film gets thicker toward the middle. Since thicker film delays the light more than thinner film, and some colors are delayed more than others, some colors will be bright and all the others not so bright for a particular thickness of film.

In the earlier photos the rainbow colors appear in rings. The outermost edge of the film is usually colorless. This is because the film there is often a single molecule thick and is thinner than a wavelength of visible light. Inside of that edge, the first ring of colors shows where the thickness of the film is first in the range for making colors.

The other inner rings are seen because the film is now two times as thick as necessary to show colors. The waves of light are now lining up where one ray has fallen more than a full wavelength behind the other.

This can continue for several rings. As the rings become more numerous, the ones caused by very thick layers overlap one another, and eventually they all combine into white light and no rings are seen. This is why you only see the rings in thin films.

A Solar-Powered Marshmallow Roaster

In this section you will learn how to make a marshmallow roaster powered by the sun. It can be made from readily available materials, and while it is probably a little safer than the traditional method of roasting marshmallows, it can still start fires and should be used only by those responsible enough to use a box of matches.

Like most of the projects in this book, it is not just a fun device, but teaches important scientific principles.

SHOPPING LIST

- Page magnifier (More technically known as a Fresnel lens—pronounced freh-NELL—it is a 7-inch by 10-inch piece of plastic that is used to magnify a page of a book to make it easier to read. It is available in drug and stationery stores, and also at www.scitoys.com.)
- Small cardboard box (The size depends on the focal length of the magnifier. The magnifier you will use focuses the sun to a small bright dot at a point 10.5 inches from the lens. This means that a box 10 inches on a side would be perfect.)
- Package of bamboo skewers to hold the marshmallows (You can also use coat-hanger wire or long fondue forks.)
- Aluminum foil

TOOLS

- ☐ Glue
- ☐ Tape
- ☐ Scissors

First, cut a hole in the box just ¼ inch smaller on each edge than the Fresnel lens.

Tape the lens to the inside of the box. The lens has a smooth side and a grooved side. The grooved side should be facing out, away from the inside of the box.

Next, glue aluminum foil to the inside of the box on all sides except the side that has the lens. This is to ensure that if the box is accidentally left in the sun, the lens will reflect off the shiny aluminum and not burn a hole in the cardboard.

The foil should be shiny side out, and it does not matter if it is wrinkled.

On the side of the box opposite the lens, cut a square hole about twice the size of a marshmallow.

On either side of the hole, cut small triangular tabs to hold the skewer. These tabs will be bent outward, and the skewer rests on them.

The photo at right shows the results of a somewhat overzealous approach to marshmallow roasting. While the outside of the marshmallow may be quite overdone, you can see from its drooping position that the inside is a warm, creamy delight.

The bright sunlight, concentrated on the highly reflective white marshmallow, is difficult to look at. Welding goggles, an inexpensive item at most hardware stores, add an extra level of excitement and awe for the participants. Very dark glasses, two pairs of dark glasses, or solar eclipse viewing glasses also work well.

A light coating of chocolate syrup or cocoa powder helps the marshmallow absorb the sunlight instead of reflecting it. This speeds up the roasting process and reduces the glare on the eyes. Those who like their marshmallows "well done" may want to first burn a small black hole in the marshmallow by holding it at the exact focus of the lens, and then expand the black spot by moving away from the focus a little bit. The black spot absorbs the sunlight very well, and the marshmallow cooks quickly.

This roaster can also be used for Vienna sausages or bite-sized pieces of hot dog.

🤖 WHY DOES IT DO THAT? 🤖

A flat plate of glass does not magnify. To magnify an image, the glass must have a curved shape, like a magnifying glass or lens does. The word *lens* comes from the Latin word for the lentil, a disk-shaped seed whose top and bottom surfaces curve outward.

The Fresnel lens used in the marshmallow roaster appears to be flat. This is because a special trick is used to make a flat magnifier.

Inside a normal lens, you can draw many rectangular areas. These areas are glass, but since they have flat edges they do not help the lens magnify. So they are not useful for the purpose of a magnifier and simply add unnecessary weight and cost to the lens.

The second part of the diagram below shows what is left if we remove the useless parts, and only keep the parts of the lens that magnify.

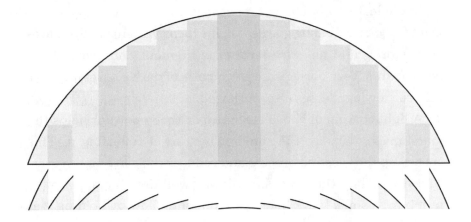

One side of our resulting *lens* is flat. But the other side has ridges with curved sides. These curved pieces of glass (or plastic in the Fresnel lens) bend the light in the same way as the original lens did. If you rub the Fresnel lens with your fingers, you can feel these ridges. (This discussion of how Fresnel lenses work is actually a simplification of what is really going on. There will be more detail later.)

Absorption

Concentrating the sunlight is only half of what is going on in the roaster. The other half is what happens when the light hits the marshmallow.

The marshmallow is white. It reflects almost all of the light that hits it. Only a small fraction of the light is absorbed. When light is absorbed by a material, it is not lost. The energy from the light moves the molecules of the marshmallow. Moving molecules is what you feel as heat.

In order to heat up the marshmallow, you had to use the very smallest dot of light from the lens, where all of the sunlight is concentrated into one tiny spot. The small fraction of the light that the marshmallow absorbs is now enough to heat up the marshmallow until it burns at that spot. But once the burned part of the marshmallow is no longer white, it no longer reflects very much light. It eventually turns black. Black objects absorb much more light than they reflect.

Now that the spot is absorbing most of the sunlight, it gets hot very quickly. If you don't move the marshmallow, it will catch fire. When you move the marshmallow closer to the lens so the circle of light from the lens is larger, it becomes less concentrated. It is still concentrated enough to roast the black spot on the marshmallow and make it bigger. By coating the marshmallow with a dark substance, like chocolate syrup or cocoa, you can speed up the heating of the marshmallow.

More About Fresnel Lenses

In the example, you simply moved the curved pieces of the lens down to lie flat. But a curve that is designed to focus light onto a point depends on the middle of the lens being farther away from the focal point than the edge. If you simply moved the pieces down, they would not focus the light to a point. The edges would focus the light to the same point as before, but as you move to the center of the lens, the focal point moves farther away, by the same amount that you moved the pieces down. Real Fresnel lenses compensate for this. The curves are made to keep the focus at the same point, regardless of how close to the center of the lens a light ray is.

Fresnel lenses are usually flat on one side. The corrections made to keep the focus at a point only work from one direction. The lenses are most commonly made to focus light in such a way that the grooved side must face the sun, and the flat side must face the focal point. If the lens is reversed, it will not focus to a sharp point. The edges will focus too close, and the center will focus too far away. This is why the grooved side of the lens must face outside the box (toward the sun).

Fun with a Big Lens

The photo on the facing page shows a Fresnel lens boiling water in a frying pan on my driveway. The pan is set well above the focal length, so it won't melt. The size of the spot of light is just a bit smaller than the frying pan, about 6 inches across.

The frame is made of 1-inch by 4-inch lumber, supported by 2-inch by 4-inch common studs. The lens itself is 40 inches across and 30 inches high. It came from a 50-inch projection television set.

Below are four U.S. pennies that were placed at the actual focus, a spot of light about the size of the hole in the penny, ¼ inch in diameter.

Three of the pennies were of the old copper alloy type (pre-1982). They contain 95 percent copper and 5 percent zinc. The penny on top was the new copper-plated zinc type. It is 97.6 percent zinc (all in the center) and 2.4 percent copper (all in the plating). Zinc has a lower melting point than brass. Much of the zinc actually burned away, leaving the pitted surface you can see in the photo. The copper plating melted and dissolved into the zinc, making the bright gold-colored brass lump that joins the other pennies together. I then moved the focus to the bottom penny and melted the hole in it. The entire procedure took only three or four seconds.

A Solar Hot Dog Cooker

In this section you will learn how to make a powerful solar concentrator that can cook four or five hot dogs in minutes.

The Solar Hot Dog Cooker is made out of a thin (⅛-inch-thick) plastic mirror that can be found at plastic shops and glass stores (although it may have to be special ordered at some stores). The plastic is bent into the shape of a parabola, so that the sun's rays are collected over an 8-square-foot area and focused in a thin line. The hot dogs are roasted on a spit placed at the focus, and turned every once in a while to prevent them from burning.

SHOPPING LIST

- ⊃ 2 pieces of plywood, ½ inch thick, 2 feet wide, and 4 feet long
- ⊃ 2 pieces of two-by-four lumber, 1½ inches thick, 3½ inches wide, and 8 feet long
- ⊃ 16 wood screws, 2 inches long
- ⊃ Stiff steel wire, 3 feet long
- ⊃ 92 small nails or wooden pegs, about an inch long
- ⊃ Plastic mirror, ⅛ inch thick, 2 feet wide, 5½ or 6 feet long (5½ feet long might work better)

TOOLS

- ▢ Drill and a bit that matches the diameter of the 92 small nails or pegs (A larger bit, over 1 inch wide, is needed for the food hole.)
- ▢ Tape measure
- ▢ Carpenter's square
- ▢ Screwdriver
- ▢ Sunglasses

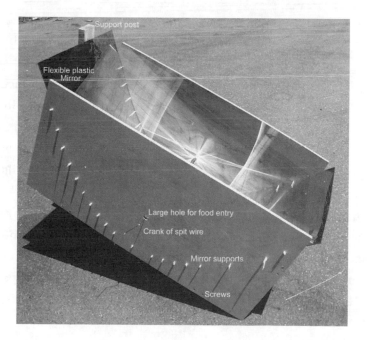

Assembly

Place the two sheets of plywood together, one on top of the other. Using a tape measure and a carpenter's square, mark off where the holes will be drilled for the mirror supports (the 92 small nails or pegs).

All holes must be drilled completely through both sheets of plywood. The holes should be drilled according to the following table:

INCHES FROM LEFT	INCHES FROM BOTTOM
0	22.16
2	18.94
4	16.00
6	13.34
8	10.96
10	8.86
12	7.04
14	5.50
16	4.24
18	3.26
20	2.56
24	2.00
28	2.56
30	3.26
32	4.24
34	5.50
36	7.04
38	8.86
40	10.96
42	13.34
44	16.00
46	18.94
48	22.16

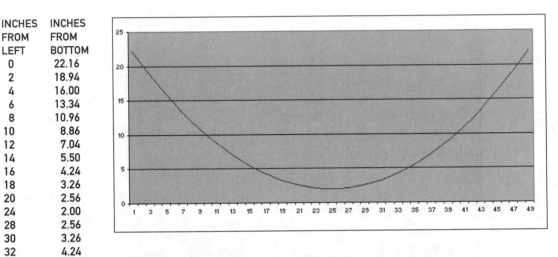

Next, drill a set of holes above the first set, about a third of an inch above the first set of holes. The first set of holes will eventually have 23 of the small nails placed in each side to hold the mirror up. The second row will also have 23 small nails pushed in, this time to hold the mirror in place from above. The exact spacing is not critical, but you don't want them too close together or the top nails will hit the mirror instead of resting on top of the mirror.

Drill eight holes for the screws that will hold the two-by-four lumber in place. The holes are ¾ inch from the edges of the plywood. On the left and right, drill a pair 15 inches from the bottom and 13 inches from the bottom. At the bottom, drill a pair 10 and 12 inches from the left, and the last pair is 36 and 38 inches from the left.

The focus of the parabola is 9.14 inches from the bottom and 24 inches from the left. Drill a hole that is the same diameter as the spit wire, or a little bit larger. This hole should go through both sheets of plywood.

Just above one of the focus holes, drill a large hole in one plywood sheet, just touching the hole for the spit. This large hole will accommodate the food (hot dogs or kabobs), so it should be at least 1 inch in diameter, but 3 or 4 inches would be better. The spit with the food on it will be inserted into this hole, and the spit will then drop into the much smaller hole at the focus, to keep the spit in exactly the right place.

Cut four pieces from the two-by-four lumber. Each piece should be exactly 2 feet long.

Using the 2-inch-long screws, screw the two-by-four pieces to one of the plywood sheets, centering each pair of screws in the end of each piece of two-by-four. The result should look something like the legs of a small table.

Attach the second plywood sheet to the other end of the two-by-four pieces.

The photo shows the backside of the cooker, where the two-by-four spreaders can be seen. Note also the remaining length of two-by-four is used as a support (more about that later).

Next, push 46 of the small nails into the bottom row of holes.

Now set the mirror onto the top of the cooker, and gently push it down to rest on the nails. Put a pair of nails in the center pair of holes on top of the mirror, then work your way outward, placing pairs of nails to hold the mirror down. (I used cotton-tipped wooden swabs in the picture because they photograph better than nails.)

The last step is to place a few screws in the remaining long piece of two-by-four, leaving the head of the screws sticking an inch or two out of the wood. These will act as supports to hold the cooker so it is tilted toward the sun.

The spit is formed from the 3-foot piece of wire. A coat hanger can be used, but wires that thin tend to sag in the middle when burdened by a few hot dogs. A thicker, stiffer wire is better. To make it easier to turn the food, bend the wire at one end to form a crank.

Cooking with the Sun

Carefully poke the 3-foot wire spit through the hot dogs or kabobs. Try to center the food on the spit so the food will rotate when you rotate the spit, instead of slipping to keep the heavy part down.

Insert the spit through the food hole, and insert the far end of the wire into the small focus hole in the far plywood sheet. Rest the near end of the spit in the small focus hole at the bottom of the food hole.

Align the solar cooker with the sun. Start with the cooker flat on the ground, then turn it until it is parallel with your shadow. The sun will just barely graze both of the plywood sheets when the cooker is aligned properly (as in the photos).

Next, tip one end of the cooker up until the shadow of the spit falls directly on the center nail at the bottom of the parabola. This can be clearly seen in the first photo.

Hold the remaining scrap of two-by-four up against the backside of the cooker, and mark where a screw should be placed to hold the cooker at the right elevation. Screw the screw into the two-by-four, leaving an inch or two sticking out to hold the top two-by-four spreader. If you like, the screw can be placed a little higher up, and the cooker can be adjusted to the exact angle by tilting the support backward.

When the cooker is adjusted properly, the sun will be focused on the food, making bright lines across it (sunglasses are recom-

mended at this phase). You can see the shadows of the nails on the walls of the cooker. These shadows should all cross at the focus, where the hot dogs are.

You can see the hot dogs in the mirror, highly magnified. The shadow of the hot dogs is being cast by the mirror onto the backside of the hot dogs in the photo above. In the photo below, you can clearly see the shadows of the nails, crossing at the focus of the parabola.

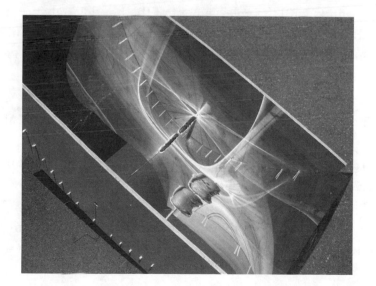

The hot dogs will start steaming in less than a minute. Turn the spit every few minutes to prevent black lines from being burned into the food (unless you like your hot dogs with black stripes). The hotdogs will be quite hot in about 10 minutes, or burned black all over in about 20 minutes.

There are a few things to notice in the above photo: (1) the shadows of the mirror supports seem to meet at the focus; (2) the shadow of the hot dog is projected onto the enlarged reflection of the hot dog; (3) the enlargement of the hot dog only occurs in its width, not its length, because the mirror is only curved in one dimension; (4) the poor hot dog has been burned to a crisp (oops!).

WHY DOES IT DO THAT?

A parabola is a shape with some interesting properties that make it perfect for cooking hot dogs with the sun.

The sun is bigger than the Earth, and very far away. This means that the sunlight that hits the Earth appears to be in parallel rays. If you had thousands of tiny mirrors, connected by hinges in a line, and you tilted each mirror so it would reflect these parallel rays onto one spot, the mirrors would line up in a parabola.

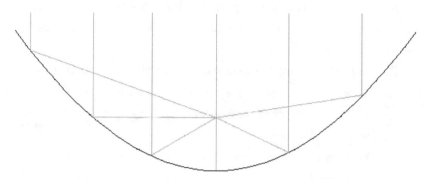

Mathematically, a parabola is defined as a set of points that are the same distance from both a point, called the *focus*, and a straight line, called the *directrix*.

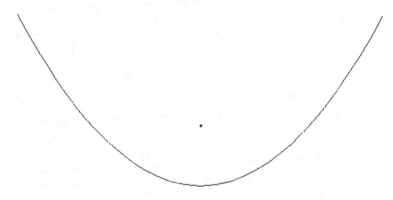

The formula for the parabola used in the solar cooker is

$$y = 0.035x^2 + 2$$

I chose this formula so the parabola would be deeply curved and would fit into the 2-foot by 4-foot plywood sheets. The focus should be close to the mirror so that as the sun moves, the focus does not move very much.

Having the focus close to the mirror is like having the fulcrum of a lever close to one end. The sun end of your lever can move a lot, while the hot dog end of your lever hardly moves at all. This means that you don't have to raise or lower the cooker very often as the sun moves.

The +2 part of the equation says that the bottom of the parabola will be 2 inches from the bottom of the plywood. This gives room for the two-by-four spreaders, and room to drill the bottom hole for the support nails.

The bottom of the parabola is called the *vertex*. The vertex is always halfway between the focus and the directrix. The distance from the vertex to the focus is

$$\frac{^1\!/0.035}{4}$$

or about 7.14 inches.

A square meter of the earth's surface gets about 1,000 watts of power from sunlight. This mirror intercepts about 8 square feet of sunlight, or about three-quarters of a square meter. This means that the cooker is the rough equivalent of a 750-watt electric stove.

A Simple Spectroscope

A spectroscope is a device that lets us find out what things are made of. It works by taking light and splitting it up into its component colors. Different elements make different colors when they glow. You can make objects and gases glow by heating them up in a flame or by passing electricity through them. The spectroscope spreads out the colors of the light, and you can identify the elements by the bright lines you see in the spectroscope.

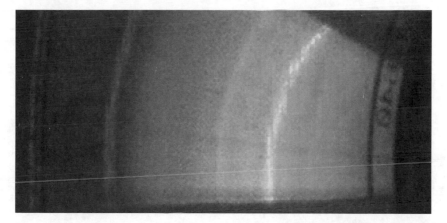

Fluorescent light

The photograph above was made using the same homemade spectroscope you will make in this project. You will see a bright green line, a bright blue-purple line, and a fainter orange line. These lines tell you that the element mercury is making some of the light. The light is coming from a fluorescent lightbulb, which works by heating up mercury until it glows. The blue-purple light from the mercury (and some invisible ultraviolet light just beyond the blue-purple line) then makes the white phosphor coating on the inside of the glass tube fluoresce bright white.

SHOPPING LIST

- CD or DVD that can be sacrificed to this project (You won't damage it, but getting it back will involve destroying your spectroscope. Old software CD-ROMs work great.)
- Cardboard box (An 8-inch cube works fine, but any size that can hold a CD or DVD will do.)
- 2 single-edged razor blades (These can be found in paint or hardware stores.)
- Small cardboard tube, the kind used as a core to wrap paper on
- Aluminum tape (found in hardware stores), or some aluminum foil and glue

TOOLS

- ☐ Cellophane tape
- ☐ Pen
- ☐ Sharp knife

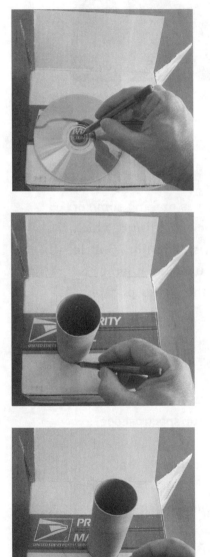

The spectroscope has three main parts: a slit made from two razor blades, a diffraction grating made from a CD, and a viewing port made from a paper tube.

To make sure that all three parts are lined up properly, you will use the CD as a measuring device, and mark the spots where the slit and the viewing port will go.

Set the CD on top of the box, about a half-inch from the left edge and close to the box's bottom, as shown in the photo. Use a pen to trace the circle inside the CD disk onto the box. This mark shows you where the paper tube will go.

Now place the paper tube on the box, centered over the circle you just drew. Draw another circle on the box by tracing the outline of the paper tube.

Move the paper tube over a little bit. A half-inch is probably fine; in the photo I placed it much farther to the right than necessary, but the aluminum tape covered up the mistake nicely. Trace another circle around the paper tube. These circles will tell us where to cut the box.

Now cut an oval out of the box with a sharp knife. The oval will allow the paper tube to enter the box at an angle.

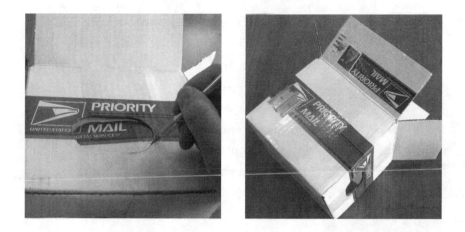

The next step is to make the slit. Turn the box one-quarter turn so that the oval you just cut is to the right. Using the CD again, draw another small circle close to the left side of the box.

The slit will be on the far left of the box. Cut a small rectangle out of the box at the height marked by the small circle you made with the CD. The rectangle should be about a half-inch wide and two inches high.

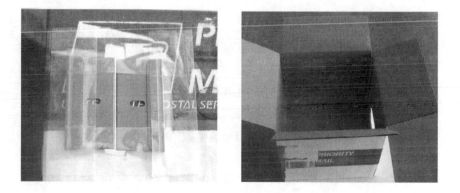

Carefully unwrap the two razor blades and set them over the rectangular hole. Make their sharp edges almost touch. Tape the razor blades to the box, being careful to leave a gap between the sharp edges that is nice and even, and not wider at the top or bottom.

Next, set the box right-side-up, with the slit toward you. Tape the CD onto the back wall of the box. The rainbow side should face

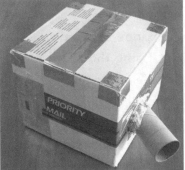

you, with the printed side touching the cardboard. The photo shows the disk a little too far to the left. The left edge of the disk should be the same distance from the left of the box as the slit is.

Now seal up any places on the box where light might leak in, using the aluminum tape. You can also use aluminum foil and glue if you don't have aluminum tape.

The last step is to use the aluminum tape to attach the paper tube. The aluminum tape will make a light-tight seal around the tube. To make sure the angle is correct, hold the slit up to a light and look through the paper tube, adjusting it until you can see the full spectrum from red to purple.

That's it! You're ready to use the spectroscope.

How to Use the Spectroscope

Hold the slit up to a source of light. An incandescent light will show a simple spectrum with no bright lines. This is because the light comes from a hot solid (the tungsten filament in the lightbulb).

Hot gases will produce light that is made up of only a few colors. The spectroscope will spread these colors out, so you can see them individually.

The photo on the facing page shows the light from a neon lightbulb. If the photo were in color, you would see that the hot neon gas is made of several colors, mostly in the red and orange parts of the spectrum.

Light is made up of waves, and each different wavelength is a different color. The neon light is showing waves of these lengths:

Neon lightbulb

- 540 nanometers (very faint) green
- 585 nanometers yellow
- 588 nanometers yellow
- 603 nanometers orange
- 607 nanometers orange
- 616 nanometers orange
- 621 nanometers red-orange
- 626 nanometers red-orange
- 633 nanometers red
- 638 nanometers red
- 640 nanometers red
- 650 nanometers red
- 660 nanometers red
- 692 nanometers red
- 703 nanometers red

Red light-emitting-diode (LED)   *Green light-emitting-diode (LED)*

A red LED makes light, but there is no hot gas, so it has a continuous spectrum.

Some green LEDs look very green. Others look more like a yellowish-green. By looking at their spectrum, as in the photo on the preceding page, you can see that the yellow-green LEDs have a lot of green, but also some yellow, orange, and red.

The LEDs have broad spectra. Their light consists of many different wavelengths. A red laser diode has a much narrower spectrum. It has only a few different wavelengths and is said to be monochromatic, meaning "one color."

Laser diode

A white LED is actually a blue LED and a phosphor. It works in a way similar to the fluorescent lightbulb, where the blue light excites the phosphors to make a white glow.

White light-emitting-diode (LED)

You will be able to see the broad spectrum in the spectroscope. The dark band in the photo is the bare spot on the CD close to the center. The spectrum is so broad that it covers the entire width of the CD.

The photos above were done using a spectroscope made from an audio CD. The photo on the facing page was done using a DVD, which has lines that are closer together. The closer lines cause the spectrum to spread out a little more.

The spectrum is not quite spread enough to show the two orange lines as separate lines. A diffraction grating with finer lines would show finer distinctions, allowing us to distinguish elements better.

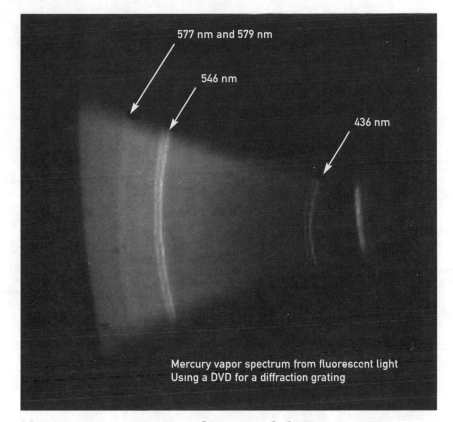

Mercury vapor spectrum in fluorescent light

A Spectroscope Made from a Cereal Box

A CD is an example of a *reflection* diffraction grating. But you can also find transmission diffraction gratings, which you look *through*, instead of look *at*.

A fairly common device using a transmission grating is a pair of paper "rainbow glasses." These are often sold at carnivals or fairs, or at fireworks displays. (www.scitoys.com also has them for sale.)

These glasses typically make rainbows up and down as well as left and right.

Making a spectroscope with rainbow glasses is very simple. You use most of the same materials and techniques as you did

for the CD spectroscope. You need a cereal box, some aluminum tape, two single-edged razor blades, and a pair of rainbow glasses. You can substitute aluminum foil and glue for the aluminum tape.

You will only need one side of the rainbow glasses, or you can use both sides and make two spectroscopes. Cut the plastic material from the glasses, leaving a paper border to keep it easy to handle.

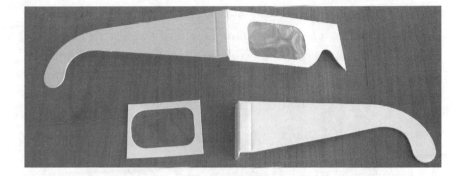

Cut a hole in the top of the cereal box on the right side, just big enough for the plastic window of the rainbow glasses. The hole should be smaller than the paper border around the plastic so the plastic doesn't fall into the box.

Tape the plastic window over the hole, and seal out any extra light with the aluminum tape.

Cut a thin rectangle out of the bottom of the cereal box, opposite the plastic window. This is where the slit will be. Seal all around the slit with aluminum tape to keep out stray light.

That's it! You now have another spectroscope!

Hold the slit up to a fluorescent lightbulb and look into the plastic window. You will see something like the photo below.

The photo above was taken using a very fast shutter speed, since the image was quite bright. The fast shutter speed only captures the primary rainbows—those closest to the slit.

However, there are also secondary rainbows farther from the slit, and they are spread out more, allowing you to more easily see the bright lines in the spectrum. The second photo used a slower shutter speed and shows the secondary spectra while washing out the primary.

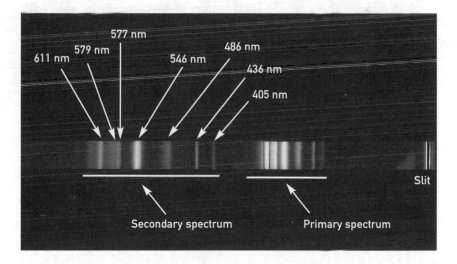

The photo above was taken from about 20 feet away from a bright (90-watt) fluorescent bulb. The lines are bright and nicely separated. With a narrower slit, I suspect the green line would resolve into two lines, and the orange line would become five or six separate lines.

I have digitally cropped out the extra spectra above and below the horizontal band.

Higher Resolution

You can get high-efficiency holographic diffraction gratings from scientific supply stores, or from www.scitoys.com. With 1,000 lines per millimeter, these gratings separate the spectral lines well, and the lines are still quite bright.

I replaced the grating in the cereal box spectroscope with one of these high-resolution gratings, and you can now clearly see several lines in the yellow and orange that were smeared together in the earlier version.

The Polariscope

In this section you will build a device called the Polariscope. As you can see from the photograph on the facing page, the Polariscope creates beautiful patterns, somewhat like a kaleidoscope, but by an entirely different mechanism.

In fact, the Polariscope is made mostly of transparent plastic, and none of the fascinating colors are visible in any of the parts

out of which it is made. The colors only
appear when the device is finally assem-
bled and you look through it. Like a kaleidoscope, when you shake the Polariscope
you get a new pattern and new colors; the
view is never the same twice.

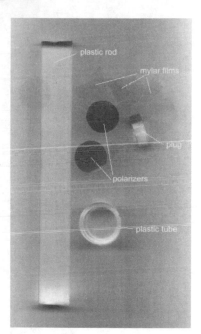

Notice that the pieces are basically colorless and clear, with the exception of the
polarizers.

SHOPPING LIST

- 8-inch long clear plastic rod, 1 inch in
  diameter. (This type of rod [called Lucite or
  Plexiglas] is available in plastics shops.)
- Piece of clear plastic tubing made of the
  same material as the rod. (It has a 1-inch inside diameter, is 1¼
  inch long, and the walls are ⅛ inch thick.)
- Polarizers (These are circles cut from Polaroid film. They can also
  be from cheap sunglasses or from old or broken LCD displays,
  such as from a watch or a small video game. You can also order
  them from scientific supply houses.)

- Cut-up pieces of Mylar film (This is available from a plastics shop. These are cut into triangles, though you can get creative with the shapes if you like.)
- Piece of plastic rod, about ⅓ inch thick, to form a plug to keep all Mylar film and the polarizers in place

TOOLS

- Hacksaw
- Sandpaper
- Polishing compound

Assembly

The plastics shop can cut and polish the ends of the rod and plug for you, or you can do it yourself. You can easily cut the rod with a fine-toothed saw, such as a hacksaw, and easily polish it with successive grades of sandpaper, up to 600 grit.

After the 600-grit paper, use polishing compound on a soft cloth to make the surface transparent. Polishing compound can be found at a plastics shop, or in any good hardware store.

Then assemble the pieces. Fit the rod into the tube, about a third of an inch in. Then place the first polarizer into the tube on top of the rod. Next pour in the Mylar pieces. Then add the second polarizer, and finally the plug.

Do not press the plug hard against the films, since they must be free to move about easily as you shake the Polariscope.

Using the Polariscope

To view through the Polariscope, hold the end with the polarizers toward the light and put your eye up to the other end. Notice that the colors change as you shake the Polariscope or turn it on its axis.

Also notice that the colors are reflected in the walls of the tube, making loops and curves out of the straight edges.

WHY DOES IT DO THAT?

The Polariscope makes use of a special property of light called *polarization*. To understand polarization, it helps to have a little background on electromagnetic waves, such as light, heat, radio waves, X-rays, and gamma rays.

Electromagnetic waves are generated when a charged object is moved. The faster the charged object moves, the more energy goes into the production of the waves, and the waves are higher in frequency.

The lowest frequency electromagnetic waves are radio waves. These can be generated by moving electrons in wires. Infrared radiation can be generated by the motion of molecules bouncing around as they are heated. As a material gets hotter, the molecules (and their electrons) move faster, and the frequency of the infrared light generated gets higher.

Visible light can be generated by moving electrons within an atom. Gamma rays can be generated by moving charges in the nucleus of an atom, or by destroying charged particles by colliding them with their antimatter counterparts.

Generating Electromagnetic Waves

It is easy to generate low-frequency radio waves using a coil of wire, such as the type in electric motors. When you connect a coil of wire to a battery, electrons flow from the battery into the coil of wire.

Moving electrons are what create magnetic fields. Electricity and magnetism can be thought of as two aspects of the same thing. Moving electrons create magnetic fields, and moving magnetic fields cause electrons to move. As the electrons move into the coil of wire, they create a magnetic field. The magnetic field starts out small and builds up as the flow of electrons builds up. At some point the current in the wire reaches its maximum (determined by the voltage of the battery and the resistance of the circuit), and the magnetic field also reaches its maximum.

If you disconnect the battery, the flow of electrons is no longer driven by the chemical forces in the battery. The magnetic field, no longer supported by the battery, begins to collapse. But since a collapsing magnetic field is a moving magnetic field, it causes the electrons in the wire to move. The electrons are pushed by the magnetic field in the same direction that the battery used to be pushing them.

Eventually the magnetic field completely collapses, and the electrons stop moving.

You know from playing with magnets that a magnetic field can affect things at a distance. You also know from playing with static electricity that charged objects can also affect things at a distance—rub a balloon on a cat and watch the cat fur rise as it is attracted to the balloon.

You have seen that as electrons start flowing in a wire, a magnetic field starts building up. You have also seen that as a magnetic field collapses it causes electrons to flow in the wire. The magnetic field and the electric field can thus be made to oscillate back and forth. As one collapses, it builds up the other.

Polarization of Electromagnetic Waves

Magnetic fields have a north pole and a south pole. Electric fields have a positive and a negative. You can picture the two fields as the height and width of a balloon as you push down on it with your hand. As the height of the balloon collapses, the width of the balloon expands. The height is like the magnetic field, and the width is like the electric field. The two fields are always at right angles to one another. This is important to understanding polarization.

Electromagnetic waves are oscillating magnetic and electric fields, at right angles to one another. They move at the speed of light (light is, after all, an electromagnetic field). Most light that we encounter is randomly oriented with respect to the oscillating fields. Since the fields are created by moving electrons, and the electrons could be moving in any direction, the fields (electric and magnetic) could be at any angle.

If something causes the electrons to all move in the same direction, then the fields would all be oriented in one direction. You would say the fields are polarized; they all have their north and south poles pointing the same way.

In your wire, the electrons are all moving in the same direction, from one end of the wire to the other. The electric and magnetic fields produced by the moving electrons are all produced in the same orientation, and the waves are polarized. Light from the sun or a candle is caused by electrons moving in all different directions, so the fields are all randomly oriented. Their light is not polarized.

The easiest way to produce polarized light from the sun or a candle is to reflect it off of a flat surface at a shallow angle. Sunlight glancing off of water into your eyes is polarized if it comes to you from a shallow enough angle. The electric field becomes oriented up and down, and the magnetic field is thus left and right.

Polarizing Filters

Some materials have their atoms aligned in rows that only allow light waves to pass through them if the electric field is parallel to the rows of atoms. Like passing a piece of paper through a comb, only the paper that is lined up with the tines of the comb can get through. Such a material is called a *polarizing filter*.

Some sunglasses are made with polarizing filters in each lens. The filter is oriented to be at 90 degrees to the light reflected off of horizontal surfaces like the water in our previous example. These filters block the light that was reflected at shallow angles, thus reducing the glare from the water. You can see this effect more clearly by rotating the glasses so they are up and down, and noticing that the glare is not reduced nearly as much as it was when the glasses were on your face in their normal position.

Rotating the Plane of Polarization

Other materials can affect polarized light in a different way. If the molecules of a transparent substance are not symmetrical, they will have more electrons on one side of the molecule than on the other. Remember that electrons and electromagnetic waves are closely related, and they affect one another. The lopsided molecules can rotate the plane of polarization of the light as it passes through the material.

The amount of rotation depends on how many wavelengths of material the light passes through. The thicker the material, the more the light is rotated. Also, the higher the frequency of light, the more the light is rotated (since there are more wavelengths in the same thickness). The Mylar pieces in the Polariscope can rotate the plane of polarization of the light.

How the Polariscope Works

In the Polariscope, light enters first through the outer polarizing filter. This polarized light then gets rotated when it passes through the Mylar. This rotated light now tries to get through the second polarizing filter, but since it has been rotated, it gets blocked.

But white light contains lots of colors. Different frequencies of light (different colors) are rotated by different amounts. Some colors will thus be rotated enough to get through the second polarizer, while other colors will be blocked.

As the light goes through different thicknesses of Mylar, it gets rotated by different amounts. Thus different colors will be blocked by different thicknesses of Mylar.

This is the magic of the Polariscope. Clear pieces of plastic can affect the colors of light that are allowed to pass though the filters. You see colors, even though there is nothing colored in the device itself.

Index